Cisco IOS 12.0 Switching services

Cisco Systems, Inc.

Copyright © 1999 Cisco Systems, Inc.

Cisco Press logo is a trademark of Cisco Systems, Inc.

Published by:
Cisco Press
201 West 103rd Street
Indianapolis, IN 46290 USA

Printed in the United States of America 1 2 3 4 5 6 7 8 9 0

Library of Congress Cataloging-in-Publication Number 99-60809

ISBN: 1-57870-157-0

Warning and Disclaimer

This book is designed to provide information about *Cisco IOS 12.0 Switching Services*. Every effort has been made to make this book as complete and as accurate as possible, but no warranty or fitness is implied.

The information is provided on an "as is" basis. The author, Cisco Press, and Cisco Systems, Inc. shall have neither liability nor responsibility to any person or entity with respect to any loss or damages arising from the information contained in this book or from the use of the discs or programs that may accompany it.

The opinions expressed in this book belong to the author and are not necessarily those of Cisco Systems, Inc.

Trademark Acknowledgments

All terms mentioned in this book that are known to be trademarks or service marks have been appropriately capitalized. Cisco Press or Cisco Systems, Inc. cannot attest to the accuracy of this information. Use of a term in this book should not be regarded as affecting the validity of any trademark or service mark.

Feedback Information

At Cisco Press, our goal is to create in-depth technical books of the highest quality and value. Each book is crafted with care and precision, undergoing rigorous development that involves the unique expertise of members from the professional technical community.

Readers' feedback is a natural continuation of this process. If you have any comments regarding how we could improve the quality of this book, or otherwise alter it to better suit your needs, you can contact us through e-mail at ciscopress@mcp.com. Please make sure to include the book title and ISBN in your message.

We greatly appreciate your assistance.

Associate Publisher	Jim LeValley
Executive Editor	Alicia Buckley
Cisco Systems Program Manager	H. Kim Lew
Managing Editor	Patrick Kanouse
Acquisitions Editor	Tracy Hughes
Copy Editor	Kitty Wilson Jarrett
Team Coordinator	Amy Lewis
Book Designer	Scott Cook
Cover Designer	Karen Ruggles
Production Team	Lisa England
	Christy Lemasters
	Elise Walter
Indexer	Craig Small

Acknowledgments

The Cisco IOS Reference Library is the result of collaborative efforts of many Cisco technical writers and editors over the years. This bookset represents the continuing development and integration of user documentation for the ever-increasing set of Cisco IOS networking features and functionality.

The current team of Cisco IOS technical writers and editors includes Katherine Anderson, Hugh Bussell, Melanie Cheng, Christy Choate, Sue Cross, Meredith Fisher, Tina Fox, Sheryl Kelly, Marsha Kinnear, Doug MacBeth, Lavanya Mandavilli, Mary Mangone, Andy Mann, Bob Marburg, Greg McMillan, Madhu Mitra, Vicki Payne, Jeremy Pollock, Patricia Rohrs, Teresa Oliver Schick, Wink Schuetz, Grace Tai, Brian Taylor, Jamianne Von-Prudelle, and Amanda Worthington.

The writing team wants to acknowledge the many engineering, customer support, and marketing subject-matter experts for their participation in reviewing draft documents and, in many cases, providing source material from which this bookset was developed.

Contents at a Glance

Table of Contents

Cisco IOS Switching Services Overview

Cisco IOS 12.0 Switching Services provides guidelines for configuring switching paths and routing between virtual local-area networks (VLANs) with the Cisco IOS software.

This guide is intended for the network administrator who designs and implements router-based internetworks and needs to incorporate switching, NetFlow accounting, or routing between VLANS into the network. It presents a set of general guidelines for configuring switching of various protocols, NetFlow accounting, routing between VLANs, and local-area network (LAN) emulation. The objective of this guide is to provide you with the information you need to configure any of these features.

You should know how to configure a Cisco router and should be familiar with the protocols and media that your routers are configured to support. Knowledge of basic network topology is essential.

Document Organization

This document comprises seven parts, each focusing on a different aspect of switching within Cisco IOS software. Each part begins with a brief technology overview and follows with the corresponding configuration guidelines for that technology or set of features. This document contains these parts:

- Cisco IOS Switching Paths – Provides an overview of basic routing and switching processes. It describes switching paths available in Cisco IOS software. Configuration guidelines are provided for configuring and managing fast switching of various protocols.

- Cisco Express Forwarding – Provides an overview of Cisco Express Forwarding (CEF), the advanced Layer 3 IP switching technology that optimizes performance and scalability in networks with large and dynamic traffic patterns. Guidelines for configuring and managing CEF are provided.

- NetFlow Switching – Provides an overview of the NetFlow switching technology and describes the NetFlow accounting features. Guidelines for configuring and managing NetFlow switching are provided.

- Tag Switching – Provides an overview of Tag Switching, the switching technology that combines the performance of Layer 2 switching with the scalability of Layer 3 routing. Guidelines for configuring and managing Tag Switching are provided.

- Multilayer Switching – Provides an overview of Multilayer Switching. Multilayer Switching provides high-performance Layer 3 switching for the Catalyst 5000 series LAN switches working in conjunction with Cisco routers. Guidelines for configuring and managing Multilayer Switching on Cisco routers are provided.

- Multicast Distributed Switching – Provides an overview of Multicast Distributed Switching (MDS). MDS performs distributed switching of multicast packets in the line cards of Route Switch Processor (RSP)-based platforms. Guidelines for configuring and managing MDS are provided.

- Virtual LANs – Provides an overview of VLANs. Guidelines for configuring routing between VLANs using the Inter-Switch Link (ISL) and IEEE 802.10 protocols for packet encapsulation follow the overview. LAN Emulation for defining VLANs in ATM networks and Multiprotocol Over ATM (MPOA) is described, along with related configuration guidelines.

About the Cisco IOS 12.0 Reference Library

The Cisco IOS 12.0 Reference Library books are Cisco documentation that describe the tasks and commands necessary to configure and maintain your Cisco IOS network.

The Cisco IOS software bookset is intended primarily for users who configure and maintain access servers and routers, but are not necessarily familiar with the tasks, the relationship between tasks, or the commands necessary to perform particular tasks.

Cisco IOS Reference Library Organization

The Cisco IOS 12.0 Reference library consists of eleven books. Each book contains technology-specific configuration chapters with corresponding command reference chapters. Each configuration chapter describes Cisco's implementation of protocols and technologies, related configuration tasks, and contains comprehensive configuration examples. Each command reference chapter complements the organization of its corresponding configuration chapter and provides complete command syntax information.

Books Available in the Cisco IOS 12.0 Reference Library

- *Cisco IOS 12.0 Solutions for Network Protocols, Volume I: IP*, 1-57870-154-6.

  This book is a comprehensive guide detailing available IP and IP routing alternatives. It describes how to implement IP addressing and IP services and how to configure support for a wide range of IP routing protocols, including BGP for ISP networks and basic and advanced IP Multicast functionality.

- *Cisco IOS 12.0 Interface Configuration*, 1-57870-156-2.

  This book is a comprehensive guide detailing how to configure physical and virtual interfaces—the two types of interfaces supported on Cisco routers. It provides readers with the most current router task and command information for their network environments and teaches how to effectively implement these techniques and commands on their networks.

- *Cisco IOS 12.0 Configuration Fundamentals*, 1-57870-155-4.

  This comprehensive guide details Cisco IOS software configuration basics. It offers thorough coverage of router and access server configuration and maintenance techniques. In addition to hands-on implementation and task instruction, this book also presents the complete syntax for router and access server commands, and individual examples for each command.

- *Cisco IOS 12.0 Wide Area Networking Solutions*, 1-57870-158-9.

  This book offers thorough, comprehensive coverage of internetworking technologies, particularly ATM, Frame Relay, SMDS, LAPB, and X.25, teaching the reader how to configure the technologies in a LAN/WAN environment.

- *Cisco IOS 12.0 Multiservice Applications*, 1-57870-159-7.

  This book shows you how to configure your router or access server to support voice, video, and broadband transmission. Cisco's voice and video support are implemented using voice packet technology. In voice packet technology, voice signals are packetized and transported in compliance with ITU-T specification H.323, which is the ITU-T specification for transmitting multimedia (voice, video, and data) across a local-area network (LAN).

- *Cisco IOS 12.0 Network Security*, 1-57870-160-0.

  This book documents security configuration from a remote site and for a central enterprise or service provider network. It describes AAA, Radius, TACACS+, and Kerberos network security features. It also explains how to encrypt data across enterprise networks. The book includes many illustrations that show configurations and functionality, along with a discussion of network security policy choices and some decision-making guidelines.

- *Cisco IOS 12.0 Quality of Service*, 1-57870-161-9.

  Cisco IOS 12.0 Quality of Service Solutions Configuration Guide is a comprehensive guide detailing available Cisco IOS quality of service (QoS) features. This book suggests benefits you can gain from implementing Cisco IOS QoS features and describes how to effectively configure and implement the various QoS features. Some of the features described in this book include Committed Access Rate (CAR), Weighted Fair Queueing (WFQ), and Weighted Random Early Detection (WRED), as well as many other features.

- *Cisco IOS 12.0 Solutions for Network Protocols, Volume II: IPX, AppleTalk, and More*, 1-57870-164-3.

  This book is a comprehensive guide detailing available network protocol alternatives. It describes how to implement various protocols in your network. This book includes documentation of the latest functionality for the IPX and AppleTalk desktop protocols as well as the following network protocols: Apollo Domain, Banyan VINES, DECNet, ISO CLNS, and XNS.

- *Cisco IOS 12.0 Bridging and IBM Network Solutions*, 1-57870-162-7.

  This book describes Cisco's support for networks in IBM and bridging environments. Support includes: transparent and source-route transparent bridging, source-route bridging (SRB), remote source-route bridging (RSRB), data link switching plus (DLS+), serial tunnel and block serial tunnel, SDLC and LLC2 parameter, IBM network media translation, downstream physical unit and SNA service point, SNA Frame Relay access support, Advanced Peer-to-Peer Networking, and native client interface architecture (NCIA).

- *Cisco IOS 12.0 Dial Solutions*, 1-57870-163-5.

  This book provides readers with real-world solutions and how to implement them on a network. Customers interested in implementing dial solutions across their network environment include remote sites dialing in to a central office, Internet Service Providers (ISPs), ISP customers at home offices, and enterprise WAN system administrators implementing dial-on-demand routing (DDR).

Book Conventions

The Cisco IOS documentation set uses the following conventions:

Convention	Description
^ or Ctrl	Represents the Control key. For example, when you read ^D or *Ctrl-D*, you should hold down the Control key while you press the D key. Keys are indicated in capital letters but are not case sensitive.
string	A string is defined as a nonquoted set of characters. For example, when setting an SNMP community string to public, do not use quotation marks around the string; otherwise, the string will include the quotation marks.

Examples use the following conventions:

Convention	Description
`screen`	Shows an example of information displayed on the screen.
`boldface screen`	Shows an example of information that you must enter.
< >	Nonprinting characters, such as passwords, appear in angled brackets.
!	Exclamation points at the beginning of a line indicate a comment line. They are also displayed by the Cisco IOS software for certain processes.
[ ]	Default responses to system prompts appear in square brackets.

The following conventions are used to attract the reader's attention:

CAUTION Means *reader be careful*. In this situation, you might do something that could result in equipment damage or loss of data.

| NOTE | Means *reader take note*. Notes contain helpful suggestions or references to materials not contained in this manual. |

| TIMESAVER | Means the *described action saves time*. You can save time by performing the action described in the paragraph. |

Within the Cisco IOS 12.0 Reference Library, the term *router* is used to refer to both access servers and routers. When a feature is supported on the access server only, the term *access server* is used.

Within examples, routers and access servers are alternately shown. These products are used only for example purposes; that is, an example that shows one product does not indicate that the other product is not supported.

Command Syntax Conventions

Command descriptions use the following conventions:

Convention	Description
boldface	Indicates commands and keywords that are entered literally as shown.
italics	Indicates arguments for which you supply values; in contexts that do not allow italics, arguments are enclosed in angle brackets (< >).
[**x**]	Keywords or arguments that appear within square brackets are optional.
{**x** | **y** | **z**}	A choice of required keywords (represented by **x**, **y**, and **z**) appears in braces separated by vertical bars. You must select one.
[**x** {**y** | **z**}]	Braces and vertical bars within square brackets indicate a required choice within an optional element. You do not need to select one. If you do, you have some required choices.

Cisco Connection Online

Cisco Connection Online (CCO) is Cisco Systems' primary, real-time support channel. Maintenance customers and partners can self-register on CCO to obtain additional information and services.

Available 24 hours a day, 7 days a week, CCO provides a wealth of standard and value-added services to Cisco's customers and business partners. CCO services include product information, product documentation, software updates, release notes, technical tips, the Bug Navigator, configuration notes, brochures, descriptions of service offerings, and download access to public and authorized files.

CCO serves a wide variety of users through two interfaces that are updated and enhanced simultaneously: a character-based version and a multimedia version that resides on the World Wide Web (WWW). The character-based CCO supports Zmodem, Kermit, Xmodem, FTP, and Internet e-mail, and it is excellent for quick access to information over lower bandwidths. The WWW version of CCO provides richly formatted documents with photographs, figures, graphics, and video, as well as hyperlinks to related information.

You can access CCO in the following ways:

- WWW: http://www.cisco.com

- WWW: http://www-europe.cisco.com

- WWW: http://www-china.cisco.com

- Telnet: cco.cisco.com

- Modem: From North America, 408 526-8070; from Europe, 33 1 64 46 40 82. Use the following terminal settings: VT100 emulation; databits: 8; parity: none; stop bits: 1; and connection rates up to 28.8 kbps.

Using Cisco IOS Software

This section provides helpful tips for understanding and configuring Cisco IOS software using the command-line interface (CLI).

Getting Help

Entering a question mark (**?**) at the system prompt displays a list of commands available for each command mode. You can also get a list of any command's associated keywords and arguments with the context-sensitive help feature.

To get help specific to a command mode, a command, a keyword, or an argument, use one of the following commands:

Command	Purpose
help	Obtains a brief description of the help system in any command mode.
*abbreviated-command-entry***?**	Obtains a list of commands that begin with a particular character string. (No space between command and question mark.)
abbreviated-command-entry<**Tab**>	Completes a partial command name.
?	Lists all commands available for a particular command mode.
command **?**	Lists a command's associated keywords. (Space between command and question mark.)
command keyword **?**	Lists a keyword's associated arguments. (Space between the keyword and question mark.)

Example: How to Find Command Options

This section provides an example of how to display syntax for a command. The syntax can consist of optional or required keywords. To display keywords for a command, enter a question mark (**?**) at the configuration prompt or after entering part of a command followed by a space. The Cisco IOS software displays a list of keywords available along with a brief description of the keywords. For example, if you were in global configuration mode, typed the command **arap**, and wanted to see all the keywords for that command, you would type **arap ?**.

Table I-1 shows examples of how you can use the question mark (**?**) to assist you in entering commands. It steps you through entering the following commands:

- `controller t1 1`
- `cas-group 1 timeslots 1-24 type e&m-fgb dtmf`

Table I-1   *How to Find Command Options*

Command	Comment
Router> **enable** Password: *<password>* Router#	Enter the **enable** command and password to access privileged EXEC commands.
	You have entered privileged EXEC mode when the prompt changes to Router#.
Router# **config terminal** Enter configuration commands, one per line. End with CNTL/Z. Router(config)#	Enter global configuration mode.
	You have entered global configuration mode when the prompt changes to Router(config)#.
Router(config)# **controller t1 ?** <0-3> Controller unit number Router(config)# **controller t1 1** Router(config-controller)#	Enter controller configuration mode by specifying the T1 controller that you want to configure using the **controller t1** global configuration command.
	Enter a **?** to display what you must enter next on the command line. In this example, you must enter a controller unit number from 0 to 3.
	You have entered controller configuration mode when the prompt changes to Router(config-controller)#.

Continues

Table I-1 *How to Find Command Options (Continued)*

Command	Comment

`Router(config-controller)# ?`
`Controller configuration commands:`
  `cablelength`   `Specify the cable length`
                        `for a DS1 link`
  `cas-group`   `Configure the specified`
                        `timeslots for CAS (Channel`
                        `Associate Signals)`
  `channel-group` `Specify the timeslots to`
                        `channel-group mapping for`
                        `an interface`
  `clock`   `Specify the clock source`
                        `for a DS1 link`
  `default`   `Set a command to its`
                        `defaults`
  `description`   `Controller specific`
                        `description`
  `ds0`   `ds0 commands`
  `exit`   `Exit from controller`
                        `configuration mode`
  `fdl`   `Specify the FDL standard`
                        `for a DS1 data link`
  `framing`   `Specify the type of`
                        `Framing on a DS1 link`
  `help`   `Description of the`
                        `interactive help system`
  `linecode`   `Specify the line encoding`
                        `method for a DS1 link`
  `loopback`   `Put the entire T1 line into`
                        `loopback`
  `no`   `Negate a command or set its`
                        `defaults`
  `pri-group`   `Configure the specified`
                        `timeslots for PRI`
  `shutdown`   `Shut down a DS1 link (send`
                        `Blue Alarm)`
`Router(config-controller)#`
	Enter a **?** to display a list of all the controller configuration commands available for the T1 controller.

`Router(config-controller)# `**`cas-group ?`**
  `<0-23>`   `Channel number`
`Router(config-controller)# cas-group`
	Enter the command that you want to configure for the controller. In this example, the **cas-group** command is used.
	
	Enter a **?** to display what you must enter next on the command line. In this example, you must enter a channel number from 0 to 23.
	
	Because a <cr> is not displayed, it indicates that you must enter more keywords to complete the command.

Table I-1 *How to Find Command Options (Continued)*

Command	Comment

`Router(config-controller)# `**`cas-group 1 ?`**
`  timeslots     List of timeslots in the`
`                cas-group`
`Router(config-controller)# cas-group 1`
	After you enter the channel number, enter a **?** to display what you must enter next on the command line. In this example, you must enter the **timeslots** keyword.

	Because a <cr> is not displayed, it indicates that you must enter more keywords to complete the command.

`Router(config-controller)# `**`cas-group 1`**
`timeslots ?`
`  <1-24>         List of timeslots which`
`                comprise the cas-group`
`Router(config-controller)# cas-group 1`
`timeslots`
	After you enter the **timeslots** keyword, enter a **?** to display what you must enter next on the command line. In this example, you must enter a list of timeslots from 1 to 24.

	You can specify timeslot ranges (for example, 1-24), individual timeslots separated by commas (for example 1, 3, 5), or a combination of the two (for example 1-3, 8, 17-24). The 16th time slot is not specified in the command line because it is reserved for transmitting the channel signaling.
	Because a <cr> is not displayed, it indicates that you must enter more keywords to complete the command.

`Router(config-controller)# `**`cas-group 1`**
`timeslots 1-24 ?`
`  service     Specify the type of service`
`  type        Specify the type of signaling`
`Router(config-controller)# cas-group 1`
`timeslots 1-24`
	After you enter the timeslot ranges, enter a **?** to display what you must enter next on the command line. In this example, you must enter the **service** or **type** keyword.

	Because a <cr> is not displayed, it indicates that you must enter more keywords to complete the command.

`Router(config-controller)# `**`cas-group 1`**
`timeslots 1-24 type ?`
`  e&m-fgb              E & M Type II FGB`
`  e&m-fgd              E & M Type IIFGD`
`  e&m-immediate-start  E & M Immediate`
`                       Start`
`  fxs-ground-start     FXS Ground Start`
`  fxs-loop-start       FXS Loop Start`
`  sas-ground-start     SAS Ground Start`
`  sas-loop-start       SAS Loop Start`
`Router(config-controller)# cas-group 1`
`timeslots 1-24 type`
	In this example, the **type** keyword is entered. After you enter the **type** keyword, enter a **?** to display what you must enter next on the command line. In this example, you must enter one of the signaling types.

	Because a <cr> is not displayed, it indicates that you must enter more keywords to complete the command.

Continues

Table I-1 *How to Find Command Options (Continued)*

Command	Comment

`Router(config-controller)# cas-group 1`
`timeslots 1-24 type e&m-fgb ?`
`  dtmf     DTMF tone signaling`
`  mf       MF tone signaling`
`  service  Specify the type of service`
`  <cr>`
`Router(config-controller)# cas-group 1`
`timeslots 1-24 type e&m-fgb`
	In this example, the **e&m-fgb** keyword is entered. After you enter the **e&m-fgb** keyword, enter a **?** to display what you must enter next on the command line. In this example, you can enter the **dtmf, mf,** or **service** keyword to indicate the type of channel-associated signaling available for the **e&m-fgb** signaling type.
	
	Because a \<cr\> is displayed, it indicates that you can enter more keywords or press \<cr\> to complete the command.

`Router(config-controller)# cas-group 1`
`timeslots 1-24 type e&m-fgb dtmf ?`
`  dnis     DNIS addr info provisioned`
`  service  Specify the type of service`
`  <cr>`
`Router(config-controller)# cas-group 1`
`timeslots 1-24 type e&m-fgb dtmf`
	In this example, the **dtmf** keyword is entered. After you enter the **dtmf** keyword, enter a **?** to display what you must enter next on the command line. In this example, you can enter the **dnis** or **service** keyword to indicate the options available for **dtmf** tone signaling.
	
	Because a \<cr\> is displayed, it indicates that you can enter more keywords or press \<cr\> to complete the command.

`Router(config-controller)# cas-group 1`
`timeslots 1-24 type e&m-fgb dtmf`
`Router(config-controller)#`
	In this example, enter a \<cr\> to complete the command.

Understanding Command Modes

The Cisco IOS user interface is divided into many different modes. The commands available to you at any given time depend on which mode you are currently in. Entering a question mark (**?**) at the system prompt allows you to obtain a list of commands available for each command mode.

When you start a session on the router, you begin in user mode, often called EXEC mode. Only a limited subset of the commands are available in EXEC mode. In order to have access to all commands, you must enter privileged EXEC mode. Normally, you must enter a password to enter privileged EXEC mode. From privileged mode, you can enter any EXEC command or enter global configuration mode. Most of the EXEC commands are one-time commands, such as **show** commands, which show the current status of something, and **clear** commands, which clear counters or interfaces. The EXEC commands are not saved across reboots of the router.

The configuration modes allow you to make changes to the running configuration. If you later save the configuration, these commands are stored across router reboots. In order to get to the various configuration modes, you must start at global configuration mode. From global configuration mode, you can enter interface configuration mode, subinterface configuration mode, and a variety of protocol-specific modes.

ROM monitor mode is a separate mode used when the router cannot boot properly. If your router or access server does not find a valid system image when it is booting, or if its configuration file is corrupted at startup, the system might enter read-only memory (ROM) monitor mode.

Summary of Main Command Modes

Table I-2 summarizes the main command modes of the Cisco IOS software.

Table I-2 *Summary of Main Command Modes*

Command Mode	Access Method	Prompt	Exit Method
User EXEC	Log in.	`Router>`	Use the **logout** command.
Privileged EXEC	From user EXEC mode, use the **enable** EXEC command.	`Router#`	To exit back to user EXEC mode, use the **disable** command. To enter global configuration mode, use the **configure terminal** privileged EXEC command.
Global configuration	From privileged EXEC mode, use the **configure terminal** privileged EXEC command.	`Router (config)#`	To exit to privileged EXEC mode, use the **exit** or **end** command or press **Ctrl-Z**. To enter interface configuration mode, enter an **interface** configuration command.
Interface configuration	From global configuration mode, enter by specifying an interface with an **interface** command.	`Router (config-if)#`	To exit to global configuration mode, use the **exit** command. To exit to privileged EXEC mode, use the **exit** command or press **Ctrl-Z**. To enter subinterface configuration mode, specify a subinterface with the **interface** command.
Subinterface configuration	From interface configuration mode, specify a subinterface with an **interface** command.	`Router (config-subif)#`	To exit to global configuration mode, use the **exit** command. To enter privileged EXEC mode, use the **end** command or press **Ctrl-Z**.
ROM monitor	From privileged EXEC mode, use the **reload** EXEC command. Press the Break key during the first 60 seconds while the system is booting.	`>`	To exit to user EXEC mode, type **continue**.

Using the No and Default Forms of Commands

Almost every configuration command also has a **no** form. In general, use the **no** form to disable a function. Use the command without the keyword **no** to reenable a disabled function or to enable a function that is disabled by default. For example, IP routing is enabled by default. To disable IP routing, specify the **no ip routing** command and specify **ip routing** to re-enable it. The Cisco IOS software command references provide the complete syntax for the configuration commands and describe what the **no** form of a command does.

Configuration commands can also have a **default** form. The **default** form of a command returns the command setting to its default. Most commands are disabled by default, so the **default** form is the same as the **no** form. However, some commands are enabled by default and have variables set to certain default values. In these cases, the **default** command enables the command and sets variables to their default values. The Cisco IOS software command references describe what the **default** form of a command does if the command is not the same as the **no** form.

Saving Configuration Changes

Enter the **copy system:running-config nvram:startup-config** command to save your configuration changes to your startup configuration so that they will not be lost if there is a system reload or power outage. For example:

```
Router# copy system:running-config nvram:startup-config
Building configuration...
```

It might take a minute or two to save the configuration. After the configuration has been saved, the following output appears:

```
[OK]
Router#
```

On most platforms, this step saves the configuration to nonvolatile random-access memory (NVRAM). On the Class A Flash file system platforms, this step saves the configuration to the location specified by the CONFIG_FILE environment variable. The CONFIG_FILE variable defaults to NVRAM.

Cisco IOS Switching Paths

CHAPTER 1

Switching Paths Overview

This chapter describes switching paths that can be configured on Cisco IOS devices. It provides an overview of switching methods. For specific configuration information, refer to Chapter 2, "Configuring Switching Paths."

Overview of Basic Router Platform Architecture and Processes

To understand how switching works, it helps to first understand the basic router architecture and where various processes occur in the router.

Fast switching is enabled by default on all interfaces that support fast switching. If you have a situation where you need to disable fast switching and fall back to the process-switching path, understanding how various processes affect the router and where they occur will help you determine your alternatives. This understanding is especially helpful when you are troubleshooting traffic problems or need to process packets that require special handling. Some diagnostic or control resources are not compatible with fast switching or come at the expense of processing and switching efficiency. Understanding those resources can help you minimize their effect on network performance.

Figure 1-1 illustrates a possible internal configuration of a Cisco 7500 series router. In this configuration, the Cisco 7500 series router has an integrated Route/Switch Processor (RSP) and uses *route caching* to forward packets. The Cisco 7500 series router also uses Versatile Interface Processors (VIPs), a RISC-based interface processor that receives and caches routing information from the RSP. The VIP card uses the route cache to make switching decisions locally, which relieves the RSP of involvement and speeds overall throughput. This type of switching is called *distributed switching*. Multiple VIP cards can be installed in one router.

Figure 1-1 *Basic Router Architecture*

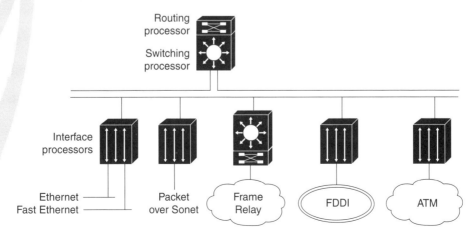

Cisco Routing and Switching Processes

The routing, or forwarding, function comprises two interrelated processes to move information in the network:

● Making a routing decision by routing

● Moving packets to the next-hop destination by switching

Cisco IOS platforms perform both routing and switching, and there are several types of each.

Routing

The routing process assesses the source and destination of traffic based on knowledge of network conditions. Routing functions identify the best path to use for moving the traffic to the destination from one or more of the router interfaces. The routing decision is based on a variation of criteria such as link speed, topological distance, and protocol. Each separate protocol maintains its own routing information.

Routing is more processing intensive and has a higher latency than switching as it determines path and next-hop considerations. The first packet routed requires a lookup in the routing table to determine the route. The route cache is populated after the first packet is routed by the route-table lookup. Subsequent traffic for the same destination is switched using the routing information stored in the route cache. Figure 1-2 illustrates the basic routing process.

Figure 1-2   *The Routing Process*

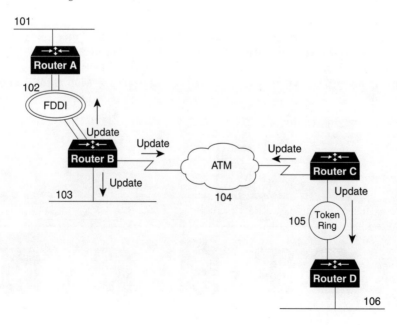

A router sends routing updates out to each of its interfaces that are configured for a particular protocol. It also receives routing updates from other attached routers. From these received updates and its knowledge of attached networks, it builds a map of the network topology.

Switching

Through the switching process, the router determines the next hop toward the destination address. Switching moves traffic from an input interface to one or more output interfaces. Switching is optimized and has a lower latency than routing because it can move packets, frames, or cells from buffer to buffer with a simpler determination of the source and destination of the traffic. It saves resources because it does not involve extra lookups. Figure 1-3 illustrates the basic switching process.

Figure 1-3   *The Switching Process*

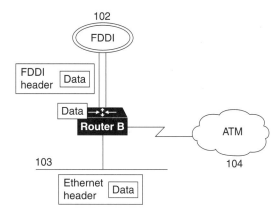

In Figure 1-3, packets are received on the Fast Ethernet interface and destined for the FDDI interface. Based on information in the packet header and destination information stored in the routing table, the router determines the destination interface. It looks in the protocol's routing table to discover the destination interface that services the destination address of the packet.

The destination address is stored in tables, such as ARP tables for IP and AARP tables for AppleTalk. If there is no entry for the destination, the router will either drop the packet (and inform the user if the protocol provides that feature), or it must discover the destination address by some other address resolution process, such as through the ARP protocol. Layer 3 IP addressing information is mapped to the Layer 2 MAC address for the next hop. Figure 1-4 illustrates the mapping that occurs to determine the next hop.

Figure 1-4 *Layer 3-to-Layer 2 Mapping*

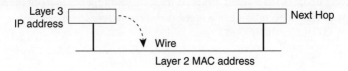

Basic Switching Paths

Basic switching paths are

● Process Switching

● Fast Switching

● Distributed Switching

● NetFlow Switching

Process Switching

In process switching, the first packet is copied to the system buffer. The router looks up the Layer 3 network address in the routing table and initializes the fast-switch cache. The frame is rewritten with the destination address and sent to the exit interface that services that destination. Subsequent packets for that destination are sent by the same switching path. The route processor computes the cyclic redundancy check (CRC).

Fast Switching

When packets are fast switched, the first packet is copied to packet memory and the destination network or host is found in the fast-switching cache. The frame is rewritten and sent to the exit interface that services the destination. Subsequent packets for the same destination use the same switching path. The interface processor computes the CRC.

Distributed Switching

Switching becomes more efficient the closer to the interface the function occurs. In distributed switching, the switching process occurs on VIP and other interface cards that support switching. Figure 1-5 illustrates the distributed switching process on the Cisco 7500 series.

Figure 1-5   *Distributed Switching on Cisco 7500 Series Routers*

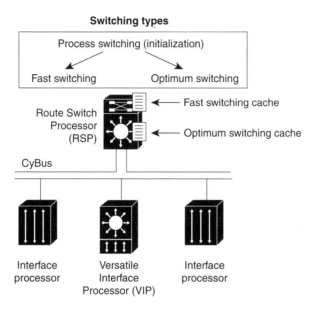

The VIP card installed in this router maintains a copy of the routing cache information needed to forward packets. Because the VIP card has the routing information it needs, it performs the switching locally, making the packet forwarding much faster. Router throughput is increased linearly based on the number of VIP cards installed in the router.

NetFlow Switching

NetFlow switching enables you to collect the data required for flexible and detailed accounting, billing, and chargeback for network and application resource utilization. Accounting data can be collected for both dedicated line and dial-access accounting. NetFlow switching over a foundation of VLAN technologies provides the benefits of switching and routing on the same platforms. NetFlow switching is supported over switched LAN or ATM backbones, allowing scalable inter-VLAN forwarding. NetFlow switching can be deployed at any location in the network as an extension to existing routing infrastructures. NetFlow switching is described in Chapter 8, "Configuring NetFlow Switching."

Platform and Switching Path Correlation

Depending on the routing platform you are using, availability and default implementations of switching paths varies. Table 1-1 shows the correlation between Cisco IOS switching paths and routing platforms.

Table 1-1   *Switching Paths on RSP-Based Routers*

Switching Path	Cisco 7200	Cisco 7500	Comments	Configuration Command
Process switching	Yes	Yes	Initializes switching caches	**no** *protocol* **route-cache**
Fast switching	Yes	Yes	Default (except for IP)	*protocol* **route-cache**
Distributed switching	No	Yes	Using second-generation VIP line cards	*protocol* **route-cache distributed**
NetFlow switching	Yes	Yes	Configurable per interface	*protocol* **route-cache flow**

Features That Affect Performance

Performance is derived from the switching mechanism you are using. Some Cisco IOS features require special handling and cannot be switched until the additional processing they require has been performed. This special handling is not processing that the interface processors can do. Because these features require additional processing, they affect switching performance. These features include

- Queuing
- Random Early Detection
- Compression
- Filtering
- Encryption
- Accounting

Queuing

Queuing occurs when network congestion occurs. When traffic is moving well within the network, packets are sent as they arrive at the interface. Cisco IOS software implements four different queuing algorithms:

- First-in, first-out (FIFO) queuing – Packets are forwarded in the same order in which they arrive at the interface.

- Priority queuing – Packets are forwarded based on an assigned priority. You can create priority lists and groups to define rules for assigning packets to priority queues.

- Custom queuing – You can control a percentage of interface bandwidth for specified traffic by creating protocol queue lists and custom queue lists.

- Weighted fair queuing – Weighted fair queuing provides automatic traffic priority management. Low-bandwidth sessions have priority over high-bandwidth sessions and high-bandwidth session are assigned weights. Weighted fair queuing is the default for interfaces slower than 2.048 Mbps.

Random Early Detection

Random early detection is designed for congestion avoidance. Traffic is prioritized based on type of service (TOS) or precedence. This feature is available on T3, OC-3, and ATM interfaces.

Compression

Depending on the protocol you are using, various compression options are available in Cisco IOS software.

Filtering

You can define access lists to control access to or from a router for a number of services. You could, for example, define an access list to prevent packets with a certain IP address from leaving a particular interface on a router. How access lists are used depends on the protocol.

Encryption

Encryption algorithms are applied to data to alter its appearance, making it incomprehensible to those who are not authorized to see the data.

Accounting

You can configure accounting features to collect network data related to resource usage. The information you collect (in the form of statistics) can be used for billing, chargeback, and planning resource usage.

CHAPTER 2

Configuring Switching Paths

This chapter describes how to configure switching paths on Cisco IOS devices. It provides configuration guidelines for switching paths and tuning guidelines. To find documentation of other commands, you can search online at www.cisco.com.

Fast Switching Configuration Task List

Fast switching allows higher throughput by switching a packet using a cache created by the initial packet sent to a particular destination. Destination addresses are stored in the high-speed cache to expedite forwarding. Routers offer better packet-transfer performance when fast switching is enabled. Fast switching is enabled by default on all interfaces that support fast switching.

Use the commands in these sections to configure appropriate fast switching features:

- Enabling AppleTalk Fast Switching
- Enabling IP Fast Switching
- Enabling Fast Switching on the Same IP Interface
- Enabling Fast Switching of IPX-Directed Broadcast Packets
- Disabling Banyan VINES Fast Switching
- Enabling Fast Switching of IPX-Directed Broadcast Packets

Fast Switching is not supported for the X.25 encapsulations.

Enabling AppleTalk Fast Switching

AppleTalk access lists are automatically fast switched. Access list fast switching improves the performance of AppleTalk traffic when access lists are defined on an interface.

Enabling IP Fast Switching

Fast switching involves the use of a high-speed switching cache for IP routing. Destination IP addresses are stored in the high-speed cache to expedite packet forwarding. In some cases, fast switching is inappropriate, such as when slow-speed serial links (64K and below) are being fed from higher-speed media such as T1 or Ethernet. In such a case, disabling fast switching can reduce the packet drop rate to some extent. Fast switching allows outgoing packets to be load balanced on a *per-destination* basis.

To enable or disable fast switching, use either of the following commands in interface configuration mode:

Command	Purpose
ip route-cache	Enables fast switching (use of a high-speed route cache for IP routing).
no ip route-cache	Disables fast switching and enable load balancing on a per-packet basis.

Enabling Fast Switching on the Same IP Interface

You can enable IP fast switching when the input and output interfaces are the same interface. This normally is not recommended, though it is useful when you have partially meshed media such as Frame Relay. You could use this feature on other interfaces, although it is not recommended because it would interfere with redirection.

Figure 2-1 illustrates a scenario where this is desirable. Router A has a data link connection identifier (DLCI) to Router B, and Router B has a DLCI to Router C. There is no DLCI between Routers A and C; traffic between them must go in and out of Router B through the same interface.

Figure 2-1   *IP Fast Switching on the Same Interface*

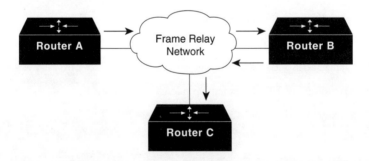

To allow IP fast switching on the same interface, use the following command in interface configuration mode:

Command	Purpose
ip route-cache same-interface	Enables the fast switching of packets out of the same interface on which they arrived.

Enabling Fast Switching of IPX-Directed Broadcast Packets

By default, Cisco IOS software switches packets that have been directed to the broadcast address. To enable fast switching of these IPX-directed broadcast packets, use the following command in global configuration mode:

Command	Purpose
ipx broadcast-fast switching	Enables fast switching of IPX directed broadcast packets.

Enabling SMDS Fast Switching

SMDS fast switching of IP, IPX, and AppleTalk packets provides faster packet transfer on serial links with speeds above 56 kbps. Use fast switching if you use high-speed, packet-switched, datagram-based WAN technologies such as Frame Relay offered by service providers.

By default, SMDS fast switching is enabled.

To re-enable fast switching, if it has been disabled, use the following commands in interface configuration mode:

Step	Command	Purpose
1	**interface** *type number*	Defines the type and unit number of the interface, and enters interface configuration mode.
2	**encapsulation smds**	Sets SMDS encapsulation.
3	**ip route-cache**	Enables the interface for IP fast switching.
4	**ipx route-cache**	Enables the interface for IPX fast switching.
5	**appletalk route-cache**	Enables the interface for AppleTalk fast switching.

Disabling Fast Switching for Troubleshooting

Fast switching uses a cache created by previous packets to achieve a higher packet throughput. Packet transfer performance is generally better when fast switching is enabled. Fast switching also provides load sharing on a per-packet basis.

By default, fast switching is enabled on all interfaces that support fast switching. However, you may want to disable fast switching to save memory space on interface cards and to help avoid congestion when high-bandwidth interfaces are writing large amounts of information to low-bandwidth interfaces. This is especially important when using rates slower than T1.

Fast switching is not supported on serial interfaces using encapsulations other than HDLC.

NOTE   Turning off fast switching increases system overhead.

For some diagnostics such as debugging and packet-level tracing, you will need to disable fast switching. If fast switching is running, you will not see packets unless they pass through the route processor. Packets would otherwise be switched on the interface. You might want to turn off fast switching temporarily and bypass the route processor while you are trying to capture information.

This section includes these topics:

- Disabling AppleTalk Fast Switching
- Disabling Banyan VINES Fast Switching
- Disabling DECnet Fast Switching
- Disabling IPX Fast Switching
- Disabling ISO CLNS Fast Switching Through the Cache
- Disabling XNS Fast Switching

Disabling AppleTalk Fast Switching

To disable AppleTalk fast switching on an interface, use the following command in interface configuration mode:

Command	Purpose
no appletalk route-cache	Disables AppleTalk fast switching.

Disabling Banyan VINES Fast Switching

Fast switching is enabled by default on all interfaces on which it is supported.

To disable fast switching on an interface, use the following command in interface configuration mode:

Command	Purpose
no vines route-cache	Disables fast switching.

Disabling DECnet Fast Switching

By default, Cisco's DECnet routing software implements fast switching of DECnet packets.

To disable fast switching of DECnet packets, use the following command in interface configuration mode:

Command	Purpose
no decnet route-cache	Disables fast switching of DECnet packets on a per-interface basis.

Disabling IPX Fast Switching

To disable IPX fast switching, use the following command in interface configuration mode:

Command	Purpose
no ipx route-cache	Disables IPX fast switching.

Disabling ISO CLNS Fast Switching Through the Cache

ISO CLNS fast switching through the cache is enabled by default for all supported interfaces. To disable fast switching, use the following command in interface configuration mode:

Command	Purpose
no clns route-cache	Disables fast switching.

NOTE	The cache still exists and is used after the **no clns route-cache** interface configuration command is used; the software just does not do fast switching through the cache.

Disabling XNS Fast Switching

To disable XNS fast switching on an interface, use the following command in interface configuration mode:

Command	Purpose
no xns route-cache	Disables XNS fast switching.

Controlling the Route Cache

The high-speed route cache used by IP fast switching is invalidated when the IP routing table changes. By default, the invalidation of the cache is delayed slightly to avoid excessive CPU load while the routing table is changing. To control the route cache, use the appropriate commands in these sections:

● Controlling Route Cache Invalidation for IP

● Displaying System and Network Statistics

● Adjusting the Route Cache for IPX

● Padding Odd-Length IPX Packets

Controlling Route Cache Invalidation for IP

To control route cache invalidation, use the following commands in global configuration mode as needed for your network:

Command	Purpose
no ip cache-invalidate-delay	Allows immediate invalidation of the cache.
ip cache-invalidate-delay [*minimum maximum quiet threshold*]	Delays invalidation of the cache.

NOTE	This task normally should not be necessary. It should be performed only under the guidance of technical staff. Incorrect configuration can seriously degrade the performance of your router.

Displaying System and Network Statistics

You can display the contents of IP routing tables and caches. The resulting information can be used to determine resource utilization and to solve network problems.

Use the following command in privileged EXEC mode:

Command	Purpose
show ip cache [*prefix mask*] [*type number*]	Displays the routing table cache used to fast switch IP traffic.

Adjusting the Route Cache for IPX

Adjusting the route cache allows you to control the size of the route cache, reduce memory consumption, and improve router performance. You accomplish these tasks by controlling the route cache size and invalidation. The following sections describe these optional tasks:

● Controlling IPX Route Cache Size

● Controlling IPX Route Cache Invalidation

Controlling IPX Route Cache Size

You can limit the number of entries stored in the IPX route cache to free up router memory and aid router processing.

Storing too many entries in the route cache can use a significant amount of router memory, causing router processing to slow. This situation is most common on large networks that run network management applications for NetWare.

For example, if a network management station is responsible for managing all clients and servers in a very large (more than 50,000 nodes) Novell network, the routers on the local segment can become inundated with route cache entries. You can set a maximum number of route cache entries on these routers to free up router memory and aid router processing.

To set a maximum limit on the number of entries in the IPX route cache, use the following command in global configuration mode:

Command	Purpose
ipx route-cache max-size *size*	Sets a maximum limit on the number of entries in the IPX route cache.

If the route cache has more entries than the specified limit, the extra entries are not deleted. However, they may be removed if route cache invalidation is in use. See the "Controlling IPX Route Cache Invalidation" section in this chapter for more information on invalidating route cache entries.

Controlling IPX Route Cache Invalidation

You can configure the router to invalidate fast switch cache entries that are inactive. If these entries remain invalidated for 1 minute, the router purges the entries from the route cache.

Purging invalidated entries reduces the size of the route cache, reduces memory consumption, and improves router performance. Purging entries also helps ensure accurate route cache information.

You can specify the period of time that valid fast switch cache entries must be inactive before the router invalidates them. You can also specify the number of cache entries that the router can invalidate per minute.

To configure the router to invalidate fast-switch cache entries that are inactive, use the following command in global configuration mode:

Command	Purpose
ipx route-cache inactivity-timeout *period* [*rate*]	Invalidates fast switch cache entries that are inactive.

When you use the **ipx route-cache inactivity-timeout** command with the **ipx route-cache max-size** command, you can ensure a small route cache with fresh entries.

Padding Odd-Length IPX Packets

Some IPX end hosts accept only even-length Ethernet packets. If the length of a packet is odd, the packet must be padded with an extra byte so that the end host can receive it. By default, Cisco IOS pads odd-length Ethernet packets.

However, there are cases in certain topologies where non-padded Ethernet packets are being forwarded onto a remote Ethernet network. Under specific conditions, you can enable padding on intermediate media as a temporary workaround for this problem. Note that you should perform this task only under the guidance of a customer engineer or other service representative.

To enable the padding of odd-length packets, use the following commands in interface configuration mode:

Step	Command	Purpose
1	**no ipx route-cache**	Disables fast switching.
2	**ipx pad-process-switched-packets**	Enables the padding of odd-length packets.

CHAPTER 3

Cisco IOS Switching Commands

This chapter documents commands used to configure switching features in Cisco IOS software.

NOTE   Beginning with Cisco IOS Release 11.3, all commands supported on the Cisco 7500 series routers are also supported on Cisco 7000 series routers.

---

clear ip flow stats

To clear the NetFlow switching statistics, use the **clear ip flow stats** EXEC command.

> **clear ip flow stats**

Syntax Description
This command has no arguments or keywords.

Command Mode
EXEC

Usage Guidelines
This command first appeared in Cisco IOS Release 11.1 CA.

The **show ip cache flow** command displays the NetFlow switching statistics. Use the **clear ip flow stats** command to clear the NetFlow switching statistics.

Example
The following example clears the NetFlow switching statistics on the router:

```
clear ip flow stats
```

Related Commands
You can search online at www.cisco.com to find documentation of related commands.

show ip cache

encapsulation isl

Use the **encapsulation isl** subinterface configuration command to enable Inter-Switch Link (ISL). ISL is a Cisco protocol for interconnecting multiple switches and routers, and for defining VLAN topologies.

encapsulation isl *vlan-identifier*

Syntax	Description
vlan-identifier	Virtual LAN identifier. The allowed range is 1 to 1000.

Default

Disabled

Command Mode

Subinterface configuration

Usage Guidelines

This command first appeared in Cisco IOS Release 11.1.

ISL encapsulation is configurable on Fast Ethernet interfaces.

ISL encapsulation adds a 26-byte header to the beginning of the Ethernet frame. The header contains a 10-bit VLAN identifier that conveys VLAN membership identities between switches.

Example

The following example enables ISL on Fast Ethernet subinterface 2/1.20:

```
interface FastEthernet 2/1.20
 encapsulation isl 400
```

Related Commands

You can search online at www.cisco.com to find documentation of related commands.

bridge-group
debug vlan
show bridge vlan
show interfaces
show vlans

encapsulation sde

Use the **encapsulation sde** subinterface configuration command to enable IEEE 802.10 encapsulation of traffic on a specified subinterface in virtual LANs. IEEE 802.10 is a standard protocol for interconnecting multiple switches and routers and for defining VLAN topologies.

> **encapsulation sde** *said*

Syntax	Description
said	Security association identifier. This value is used as the virtual LAN identifier. The valid range is 0 through 0xFFFFFFFE.

Default

Disabled

Command Mode

Subinterface configuration

Usage Guidelines

This command first appeared in Cisco IOS Release 10.3.

SDE encapsulation is configurable only on the following interface types:

IEEE 802.10 Routing	IEEE 802.10 Transparent Bridging
• FDDI	• Ethernet
	• FDDI
	• HDLC serial
	• Transparent mode
	• Token Ring

Example

The following example enables SDE on FDDI subinterface 2/0.1 and assigns a VLAN identifier of 9999:

```
interface fddi 2/0.1
 encapsulation sde 9999
```

Related Commands

You can search online at www.cisco.com to find documentation of related commands.

bridge-group
debug vlans
show bridge vlan2222
show interfaces
show vlans

encapsulation tr-isl

Use the **encapsulation tr-isl** subinterface configuration command to enable TRISL, a Cisco proprietary protocol for interconnecting multiple routers and switches and maintaining VLAN information as traffic goes between switches.

encapsulation tr-isl trbrf-vlan *vlan-id* **bridge-num** *bridge-number*

Syntax

vlan-id

bridge-number

Description

Number identifying the VLAN.

Keyword that specifies the identification number of the bridge number on the ISL trunk. Possible values are 01 to 4095.

Command Mode

Subinterface configuration

Usage Guidelines

This command first appeared in Cisco IOS Release 11.3(4)T.

Examples

In the following example, TRISL is enabled on a Fast Ethernet interface:

```
interface FastEthernet4/0.2
 encapsulation tr-isl trbrf-vlan 999 bridge-num 14
```

Related Commands

You can search online at www.cisco.com to find documentation of related commands.

clear drip counters
clear vlan statistics
multiring
multiring trcrf-vlan
show drip
show vlan
source-bridge trcrf-vlan

ip cache-invalidate-delay

To control the invalidation rate of the IP route cache, use the **ip cache-invalidate-delay** global configuration command. To allow the IP route cache to be immediately invalidated, use the **no** form of this command.

> **ip cache-invalidate-delay** [*minimum maximum quiet threshold*]
> **no ip cache-invalidate-delay**

Syntax   Description

Syntax	Description
minimum	(Optional) Minimum time (in seconds) between invalidation request and actual invalidation. The default is 2 seconds.
maximum	(Optional) Maximum time (in seconds) between invalidation request and actual invalidation. The default is 5 seconds.
quiet	(Optional) Length of quiet period (in seconds) before invalidation.
threshold	(Optional) Maximum number of invalidation requests considered to be quiet.

Defaults

minimum = 2 seconds
maximum = 5 seconds, and 3 seconds with no more than zero invalidation requests

Command Mode

Global configuration

Usage Guidelines

This command first appeared in Cisco IOS Release 10.0.

All cache invalidation requests are honored immediately.

This command should typically not be used except under the guidance of technical support personnel. Incorrect settings can seriously degrade network performance.

The IP fast-switching and autonomous-switching features maintain a cache of IP routes for rapid access. When a packet is to be forwarded and the corresponding route is not present in the cache, the packet is process-switched and a new cache entry is built. However, when routing table changes occur (such as when a link or an interface goes down), the route cache must be flushed so that it can be rebuilt with up-to-date routing information.

This command controls how the route cache is flushed. The intent is to delay invalidation of the cache until after routing has settled down. Because route table changes tend to be clustered in a short period of time, and the cache may be flushed repeatedly, a high CPU load might be placed on the router.

When this feature is enabled, and the system requests that the route cache be flushed, the request is held for at least *minimum* seconds. Then the system determines whether the cache has been "quiet" (that is, less than *threshold* invalidation requests in the last *quiet* seconds). If the cache has been quiet, the cache is then flushed. If the cache does not become quiet within *maximum* seconds after the first request, it is flushed unconditionally.

Manipulation of these parameters trades off CPU utilization versus route convergence time. Timing of the routing protocols is not affected, but removal of stale cache entries is affected.

Example

The following example sets a minimum delay of 5 seconds, a maximum delay of 30 seconds, and a quiet threshold of no more than 5 invalidation requests in the previous 10 seconds:

```
ip cache-invalidate-delay 5 30 10 5
```

Related Commands

You can search online at www.cisco.com to find documentation of related commands.

ip route-cache
show ip cache

ip flow-cache entries

Use the **ip flow-cache entries** global configuration command to change the number of entries maintained in the NetFlow cache. Use the **no** form of this command to return to the default number of entries.

ip flow-cache entries *number*
no ip flow-cache entries

Syntax

number

Description

Number of entries to maintain in the NetFlow cache. The valid range is 1024 to 524288 entries. The default is 65536 (64 K).

Default

65536 entries (64K)

Command Mode

Global configuration

Usage Guidelines

This command first appeared in Cisco IOS Release 11.1 CA.

Normally the default size of the NetFlow cache will meet your needs. However, you can increase or decrease the number of entries maintained in the cache to meet the needs of your flow traffic rates. For environments with a high amount of flow traffic (such as an internet core router), a larger value such as 131072 (128K) is recommended. To obtain information on your flow traffic, use the **show ip cache flow** command.

The default is 64K flow cache entries. Each cache entry is approximately 64 bytes of storage. Assuming a cache with the default number of entries, approximately 4MB of DRAM would be required. Each time a new flow is taken from the free flow queue, the number of free flows is checked. If there are only a few free flows remaining, NetFlow attempts to age 30 flows using an accelerated timeout. If there is only one free flow remaining, NetFlow automatically ages 30 flows regardless of their age. The intent is to ensure free flow entries are always available.

CAUTION Cisco recommends that you do not change the NetFlow cache entries. Improper use of this command could cause network problems. To return to the default NetFlow cache entries, use the **no ip flow-cache entries** global configuration command.

Example

The following example increases the number of entries in the NetFlow cache to 131072 (128K):

```
ip flow-cache entries 131072
```

Related Commands

You can search online at www.cisco.com to find documentation of related commands.

show ip cache

ip flow-export

To enable the exporting of information in NetFlow cache entries, use the **ip flow-export** global configuration command. To disable the exporting of information, use the **no** form of this command.

ip flow-export *ip-address udp-port* [**version 1** | **version 5** [**origin-as** | **peer-as**]]
no ip flow-export

Syntax	Description
ip-address	IP address of the workstation to which you want to send the NetFlow information.
udp-port	UDP protocol-specific port number.
version 1	(Optional) Specifies that the export packet uses the version 1 format. This is the default. The version field occupies the first two bytes of the export record. The number of records stored in the datagram is a variable between 1 and 24 for version 1.
version 5	(Optional) Specifies export packet uses the version 5 format. The number of records stored in the datagram is a variable between 1 and 30 for version 5.
origin-as	(Optional) Specifies that export statistics includes the origin autonomous system (AS) for the source and destination.
peer-as	(Optional) Specifies that export statistics includes the peer AS for the source and destination.

Default

Disabled

Command Mode

Global configuration

Usage Guidelines

This command first appeared in Cisco IOS Release 11.1.

This command was modified to include the **version** keyword in Cisco IOS Release 11.1 CA.

There is a lot of information in a NetFlow cache entry. When flow switching is enabled with the **ip route-cache flow** command, you can use the **ip flow-export** command to configure the router to export the flow cache entry to a workstation when a flow expires. This feature can be useful for purposes of statistics, billing, and security.

Version 5 format includes the source and destination AS addresses, source and destination prefix masks, and a sequence number. Because this change may appear on your router as a maintenance release, support for version 1 format is maintained with the **version 1** keyword.

For more information on version 1 and version 5 data format, refer to Chapter 8, "Configuring NetFlow Switching."

Examples

The following example configures the router to export the NetFlow cache entry to UDP port 125 on the workstation at 134.22.23.7 when the flow expires, using version 1 format:

```
ip flow-export 134.22.23.7 125
```

The following example configures the router to export the NetFlow cache entry to UDP port 2048 on the workstation at 134.22.23.7 when the flow expires, using version 5 format and including the peer AS information:

```
ip flow-export 134.22.23.7 2048 version 5 peer-as
```

Related Command

You can search online at www.cisco.com to find documentation of related commands.

ip route-cache flow

ip route-cache

Use the **ip route-cache** interface configuration command to control the use of high-speed switching caches for IP routing. To disable any of these switching modes, use the **no** form of this command.

> **ip route-cache [cbus]**
> **no ip route-cache [cbus]**
>
> **ip route-cache same-interface**
> **no ip route-cache same-interface**
>
> **ip route-cache [flow]**
> **no ip route-cache [flow]**

ip route-cache distributed
no ip route-cache distributed

Syntax	Description
cbus	(Optional) Enables both autonomous switching and fast switching.
same-interface	Enables fast-switching packets back out the interface on which they arrived.
flow	(Optional) Enables the Route Switch Processor (RSP) to perform flow switching on the interface.
distributed	Enables Versatile Interface Processor (VIP) distributed switching on the interface. This feature can be enabled on Cisco 7500 series routers with an RSP and VIP controllers. If both **ip route-cache flow** and **ip route-cache distributed** are configured, the VIP does distributed flow switching. If only **ip route-cache distributed** is configured, the VIP does distributed switching.

Defaults

IP autonomous switching is disabled.
Fast switching varies by interface and media.
Distributed switching is disabled.

Command Mode

Interface configuration

Usage Guidelines

This command first appeared in Cisco IOS Release 10.0. The **distributed** keyword first appeared in Cisco IOS Release 11.2.

Using the route cache is often called *fast switching*. The route cache allows outgoing packets to be load-balanced on a *per-destination* basis.

The **ip route-cache** command with no additional keywords enables fast switching.

Cisco routers generally offer better packet transfer performance when fast switching is enabled, with one exception. On networks using slow serial links (64K and below), disabling fast switching to enable the per-packet load sharing is usually the best choice.

You can enable IP fast switching when the input and output interfaces are the same interface, using the **ip route-cache same-interface** command. This normally is not recommended, though it is useful when you have partially meshed media, such as Frame Relay. You could use this feature on other interfaces, although it is not recommended because it would interfere with redirection.

When the RSP is flow switching, it uses a flow cache instead of a destination network cache to switch IP packets. The flow cache uses source and destination network address, protocol, and source and destination port numbers to distinguish entries.

The flow caching option can also be used to allow statistics to be gathered with a finer granularity. The statistics include IP subprotocols, well-known ports, total flows, average number of packets per flow, and average flow lifetime.

On Cisco 7500 series routers with RSP and VIP controllers, the VIP hardware can be configured to switch packets received by the VIP with no per-packet intervention on the part of the RSP. When VIP distributed switching is enabled, the input VIP interface tries to switch IP packets instead of forwarding them to the RSP for switching. Distributed switching helps decrease the demand on the RSP.

Not all switching methods are available on all platforms.

Examples

The following example enables both fast switching and autonomous switching:

```
ip route-cache cbus
```

The following example disables both fast switching and autonomous switching:

```
no ip route-cache
```

The following example turns off autonomous switching only:

```
no ip route-cache cbus
```

The following example enables VIP distributed flow switching on the interface:

```
interface ethernet 0/5/0
 ip address 17.252.245.2 255.255.255.0
 ip route-cache distributed
 ip route-cache flow
```

The following example returns the system to its defaults (fast switching enabled; autonomous switching disabled):

```
ip route-cache
```

Related Commands

You can search online at www.cisco.com to find documentation of related commands.

ip cache-invalidate-delay
show ip cache

ip route-cache flow

To enable NetFlow switching for IP routing, use the **ip route-cache flow** interface configuration command. To disable NetFlow switching, use the **no** form of this command.

> **ip route-cache flow**
> **no ip route-cache flow**

Syntax Description

This command has no arguments or keywords.

Default

Disabled

Command Mode

Interface configuration

Usage Guidelines

This command first appeared in Cisco IOS Release 11.1.

NetFlow switching is a high-performance, network-layer switching path that captures as part of its switching function a rich set of traffic statistics. These traffic statistics include user, protocol, port, and type of service information that can be used for a wide variety of purposes such as network analysis and planning, accounting, and billing. To export NetFlow data, use the **ip flow-export** global configuration command.

NetFlow switching is supported on IP and IP encapsulated traffic over all interface types and encapsulations except for ISL/VLAN, ATM and Frame Relay interfaces when more than one input access control list is used on the interface, and ATM LANE.

In conventional switching at the network layer, each incoming packet is handled on an individual basis with a series of functions to perform access list checks, capture accounting data, and switch the packet. With NetFlow switching, after a flow has been identified and access list processing of the first packet in the flow has been performed, all subsequent packets are handled on a connection-oriented basis as part of the flow, where access list checks are bypassed and packet switching and statistics capture are performed in tandem.

A network flow is identified as a unidirectional stream of packets between a source and destination—both defined by a network-layer IP address and transport-layer port number. Specifically, a flow is identified as the combination of the following fields:

● Source IP Address

● Destination IP Address

- Source Port Number

- Destination Port Number

- Protocol Type

- Type of Service

- Input Interface

NetFlow switching operates by creating a flow cache that contains the information needed to switch and perform the access list check for all active flows. The NetFlow cache is built by processing the first packet of a flow through the standard fast-switching path. As a result, each flow is associated with an incoming and outgoing interface port number and with a specific security access permission and encryption policy. The cache also includes entries for traffic statistics that are updated in tandem with the switching of subsequent packets. After the NetFlow cache is created, packets identified as belonging to an existing flow can be switched based on the cached information and security access list checks bypassed. Flow information is maintained within the NetFlow cache for all active flows.

NetFlow switching is one of the available switching modes. When you configure NetFlow on an interface, the other switching modes are not used on that interface. Also, with NetFlow switching you can export data (traffic statistics) to a remote workstation for further processing.

NetFlow switching is based on identifying packet flows and performing switching and access list processing within a router. It does not involve any connection-setup protocol either between routers or to any other networking device or end station and does not require any change externally—either to the traffic or packets themselves or to any other networking device. Thus, NetFlow switching is completely transparent to the existing network, including end stations and application software and network devices such as LAN switches. Also, because NetFlow switching is performed independently on each internetworking device, it does not need to be operational on each router in the network. Network planners can selectively invoke NetFlow switching (and NetFlow data export) on a router/interface basis to gain traffic performance, control, or accounting benefits in specific network locations.

NOTE   NetFlow consumes more memory and CPU resources than other switching modes; therefore, it is important to understand the resources required on your router before enabling NetFlow.

Examples

The following example enables NetFlow switching on the interface:

```
interface ethernet 0/5/0
 ip address 17.252.245.2 255.255.255.0
 ip route-cache flow
```

The following example returns the interface to its defaults (fast switching enabled; autonomous switching disabled):

```
interface ethernet 0/5/0
 ip route-cache
```

Related Commands

You can search online at www.cisco.com to find documentation of related commands.

ip flow-export
show ip cache

show ip cache

To display the routing table cache used to fast switch IP traffic, use the **show ip cache** EXEC command.

show ip cache [*prefix mask*] [*type number*]

Syntax	Description
prefix	(Optional) Display only the entries in the cache that match the prefix and mask combination.
mask	(Optional) Display only the entries in the cache that match the prefix and mask combination.
type	(Optional) Display only the entries in the cache that match the interface type and number combination.
number	(Optional) Display only the entries in the cache that match the interface type and number combination.

Command Mode
EXEC

Usage Guidelines

This command first appeared in Cisco IOS Release 10.0. The arguments *prefix*, *mask*, *type*, and *number* first appeared in Cisco IOS Release 10.0. The **show ip cache** display shows MAC headers up to 92 bytes.

Sample Displays

The following is sample output from the **show ip cache** command:

```
Router# show ip cache

IP routing cache version 4490, 141 entries, 20772 bytes, 0 hash overflows
Minimum invalidation interval 2 seconds, maximum interval 5 seconds,
   quiet interval 3 seconds, threshold 0 requests
Invalidation rate 0 in last 7 seconds, 0 in last 3 seconds
Last full cache invalidation occurred 0:06:31 ago

Prefix/Length      Age       Interface       MAC Header
131.108.1.1/32     0:01:09   Ethernet0/0     AA000400013400000C0357430800
131.108.1.7/32     0:04:32   Ethernet0/0     00000C01281200000C0357430800
131.108.1.12/32    0:02:53   Ethernet0/0     00000C029FD000000C0357430800
131.108.2.13/32    0:06:22   Fddi2/0         00000C05A3E000000C035753AAAA0300
                                             00000800
131.108.2.160/32   0:06:12   Fddi2/0         00000C05A3E000000C035753AAAA0300
                                             00000800
131.108.3.0/24     0:00:21   Ethernet1/2     00000C026BC600000C03574D0800
131.108.4.0/24     0:02:00   Ethernet1/2     00000C026BC600000C03574D0800
131.108.5.0/24     0:00:00   Ethernet1/2     00000C04520800000C03574D0800
131.108.10.15/32   0:05:17   Ethernet1/2     00000C025FF500000C0357450800
131.108.11.7/32    0:04:08   Ethernet1/2     00000C010E3A00000C03574D0800
131.108.11.12/32   0:05:10   Ethernet0/0     00000C01281200000C0357430800
131.108.11.57/32   0:06:29   Ethernet0/0     00000C01281200000C0357430800
```

Table 3-1 describes significant fields shown in the sample output from the **show ip cache** command.

Table 3-1 *show ip cache Field Descriptions*

Field	Description
IP routing cache version	Version number of this table. This number is incremented any time the table is flushed.
entries	Number of valid entries.
bytes	Number of bytes of processor memory for valid entries.
hash overflows	Number of times autonomous switching cache overflowed.
minimum invalidation interval	Minimum time delay between cache invalidation request and actual invalidation.
maximum invalidation interval	Maximum time delay between cache invalidation request and actual invalidation.
quiet interval	Length of time between cache flush requests before the cache will be flushed.
threshold n requests	Maximum number of requests that can occur while the cache is considered quiet.
invalidation rate n in last m seconds	Number of cache invalidations during the last m seconds.

Continues

Table 3-1 *show ip cache Field Descriptions*

Field	Description
0 in last 3 seconds	Number of cache invalidation requests during the last quiet interval.
last full cache invalidation occurred *hh:mm:ss* ago	Time since last full cache invalidation was performed.
prefix/length	Network reachability information for cache entry.
age	Age of cache entry.
interface	Output interface type and number.
MAC Header	Layer 2 encapsulation information for cache entry.

The following is sample output from the **show ip cache** command with a prefix and mask specified:

```
Router# show ip cache 131.108.5.0 255.255.255.0

IP routing cache version 4490, 119 entries, 17464 bytes, 0 hash overflows
Minimum invalidation interval 2 seconds, maximum interval 5 seconds,
   quiet interval 3 seconds, threshold 0 requests
Invalidation rate 0 in last second, 0 in last 3 seconds
Last full cache invalidation occurred 0:11:56 ago

Prefix/Length      Age       Interface     MAC Header
131.108.5.0/24     0:00:34   Ethernet1/2   00000C04520800000C03574D0800
```

The following is sample output from the **show ip cache** command with an interface specified:

```
Router# show ip cache e0/2

IP routing cache version 4490, 141 entries, 20772 bytes, 0 hash overflows
Minimum invalidation interval 2 seconds, maximum interval 5 seconds,
   quiet interval 3 seconds, threshold 0 requests
Invalidation rate 0 in last second, 0 in last 3 seconds
Last full cache invalidation occurred 0:06:31 ago

Prefix/Length      Age       Interface     MAC Header
131.108.10.15/32   0:05:17   Ethernet0/2   00000C025FF500000C0357450800
```

show ip cache flow

To display a summary of the NetFlow switching statistics, use the **show ip cache flow** EXEC command.

show ip cache flow

Syntax Description

This command has no arguments or keywords.

Command Mode

EXEC

Usage Guidelines

This command first appeared in Cisco IOS Release 11.1.

This command was modified to update the display with the latest information in Cisco IOS Release 11.1 CA.

Sample Display

The following is a sample output from the **show ip cache flow** command:

```
Router# show ip cache flow
IP packet size distribution (12718M total packets):
   1-32    64    96   128   160   192   224   256   288   320   352   384   416   448   480
  .000  .554  .042  .017  .015  .009  .009  .009  .013  .030  .006  .007  .005  .004  .004

   512   544   576  1024  1536  2048  2560  3072  3584  4096  4608
  .003  .007  .139  .019  .098  .000  .000  .000  .000  .000  .000

IP Flow Switching Cache, 4456448 bytes
  65509 active, 27 inactive, 820628747 added
  955454490 ager polls, 0 flow alloc failures
  Exporting flows to 1.1.15.1 (2057)
  820563238 flows exported in 34485239 udp datagrams, 0 failed
  last clearing of statistics 00:00:03

Protocol         Total  Flows   Packets Bytes  Packets Active(Sec) Idle(Sec)
--------         Flows  /Sec    /Flow  /Pkt    /Sec    /Flow       /Flow
TCP-Telnet      2656855  4.3        86    78    372.3      49.6       27.6
TCP-FTP         5900082  9.5         9    71     86.8      11.4       33.1
TCP-FTPD        3200453  5.1       193   461   1006.3      45.8       33.4
TCP-WWW       546778274 887.3       12   325  11170.8       8.0       32.3
TCP-SMTP       25536863 41.4        21   283    876.5      10.9       31.3
TCP-X            116391  0.1       231   269     43.8      68.2       27.3
TCP-BGP           24520  0.0        28   216      1.1      26.2       39.0
TCP-Frag          56847  0.0        24   952      2.2      13.1       33.2
TCP-other      49148540 79.7        47   338   3752.6      30.7       32.2
UDP-DNS       117240379 190.2        3   112    570.8       7.5       34.7
UDP-NTP         9378269 15.2         1    76     16.2       2.2       38.7
UDP-TFTP           8077  0.0         3    62      0.0       9.7       33.2
UDP-Frag          51161  0.0        14   322      1.2      11.0       39.4
UDP-other      45502422 73.8        30   174   2272.7       8.5       37.8
ICMP           14837957 24.0         5   224    125.8      12.1       34.3
IGMP              40916  0.0       170   207     11.3     197.3       13.5
IPINIP             3988  0.0     48713   393    315.2     644.2       19.6
GRE                3838  0.0        79   101      0.4      47.3       25.9
IP-other          77406  0.1        47   259      5.9      52.4       27.0
Total:        820563238 1331.7      15   304  20633.0       9.8       33.0
```

```
SrcIf      SrcIPaddress    DstIf    DstIPaddress    Pr SrcP DstP Pkts B/Pk Active
Fd0/0      80.0.0.3        Hs1/0    200.1.9.1       06 0621 0052    7   87   5.9
Fd0/0      80.0.0.3        Hs1/0    200.1.8.1       06 0620 0052    7   87   1.8
Hs1/0      200.0.0.3       Fd0/0    80.1.10.1       06 0052 0621    6   58   1.8
Hs1/0      200.0.0.3       Fd0/0    80.1.1.1        06 0052 0620    5   62   5.9
Fd0/0      80.0.0.3        Hs1/0    200.1.3.1       06 0723 0052   16   68   0.3
HS1/0      200.0.0.3       Fd0/0    80.1.2.1        06 0052 0726    6   58  11.8
Fd0/0      80.0.0.3        Hs1/0    200.1.5.1       06 0726 0052    6   96   0.3
Hs1/0      200.0.0.3       Fd0/0    80.1.4.1        06 0052 0442    3   76   0.3
Hs1/0      200.0.0.3       Fd0/0    80.1.7.1        06 0052 D381   11 1171   0.6
```

Table 3-2 describes the fields in the packet size distribution lines of the output from the **show ip cache flow** command.

Table 3-2 *Packet Size Distribution Field Descriptions*

Field	Description
IP packet size distribution	The two lines below this banner show the percentage distribution of packets by size range. In this display, 55.4% of the packets fall in the size range 33 to 64 bytes.

Table 3-3 describes the fields in the flow switching cache lines of the output from the **show ip cache flow** command.

Table 3-3 *Flow Switching Cache Display Field Descriptions*

Field	Description
bytes	Number of of bytes of memory the NetFlow cache uses.
active	Number of active flows in the NetFlow cache at the time this command was entered.
inactive	Number of flow buffers allocated in the NetFlow cache, but are not currently assigned to a specific flow at the time this command was entered.
added	Number of flows created since the start of the summary period.
ager polls	Number of times the NetFlow code looked at the cache to expire entries (used by Cisco for diagnostics only).
flow alloc failures	Number of times the NetFlow code tried to allocate a flow but could not.
exporting flows	IP address and UDP port number of the workstation to which flows are exported.
flows exported in udp datagrams	Total number of flows exported and the total number of UDP datagrams used to export the flows to the workstation.
failed	Number of flows that could not be exported by the router because of output interface limitations.
last clearing of statistics	Standard time output (hh:mm:ss) since the **clear ip flow stats** command was executed. This time output changes to hours and days after the time exceeds 24 hours.

Part
I

Command Reference

Table 3-4 describes the fields in the activity-by-protocol lines of the output from the **show ip cache flow** command.

Table 3-4   *Activity-by-Protocol Display Field Descriptions*

Field	Description
Protocol	IP protocol and the "well known" port number as described in RFC 1340.
Total Flows	Number of flows for this protocol since the last time statistics were cleared.
Flows/Sec	Average number of flows for this protocol seen per second; equal to total flows/number of seconds for this summary period.
Packets/Flow	Average number of packets observed for the flows seen for this protocol. Equal to Total Packets for this protocol or number of flows for this protocol for this summary period.
Bytes/Pkt	Average number of bytes observed for the packets seen for this protocol (total bytes for this protocol or the total number of packet for this protocol for this summary period).
Packets/Sec	Average number of packets for this protocol per second (total packets for this protocol) or the total number of seconds for this summary period).
Active(Sec)/Flow	Sum of all the seconds from the first packet to the last packet of an expired flow (for example, TCP FIN, time-out, and so forth) in seconds or total flows for this protocol for this summary period.
Idle(Sec)/Flow	Sum of all the seconds from the last packet seen in each nonexpired flow for this protocol until the time this command was entered, in seconds or total flows for this protocol for this summary period.

Table 3-5 describes the fields in the current flow lines of the output from the **show ip cache flow** command.

Table 3-5   *Current Flow Display Field Descriptions*

Field	Description
SrcIf	Internal port name for the source interface.
SrcIPaddress	Source IP address for this flow.
DstIf	Router's internal port name for the destination interface.
DstIPaddress	Destination IP address for this flow.
Pr	IP protocol; for example, 6=TCP, 17=UDP, … as defined in RFC 1340.
SrcP	Source port address, TCP/UDP "well known" port number, as defined in RFC 1340.
DstP	Destination port address, TCP/UDP "well known" port number, as defined in RFC 1340.
Pkts	Number of packets observed for this flow.
B/Pkt	Average observed number of bytes per packet for this flow.
Active	Number of seconds between first and last packet of a flow.

Related Commands

You can search online at www.cisco.com to find documentation of related commands.

ip route-cache
clear ip flow stats

PART II

Cisco Express Forwarding

CHAPTER 4

Cisco Express Forwarding Overview

Cisco Express Forwarding (CEF) is an advanced Layer 3 IP switching technology. CEF optimizes network performance and scalability for networks with large and dynamic traffic patterns, such as the Internet, on networks characterized by intensive Web-based applications, or interactive sessions.

CEF offers these benefits:

- Improved performance—CEF is less CPU-intensive than fast-switching route caching. More CPU processing power can be dedicated to Layer 3 services such as quality of service (QoS) and encryption.

- Scalability—CEF offers full switching capacity at each line card when dCEF mode is active.

- Resilience—CEF offers unprecedented level of switching consistency and stability in large dynamic networks. In dynamic networks, fast switching cache entries are frequently invalidated due to routing changes. These changes can cause traffic to be process switched using the routing table, rather than fast switched using the route cache. Because the FIB lookup table contains all known routes that exist in the routing table, it eliminates route cache maintenance and the fast switch/ process switch forwarding scenario. CEF can switch traffic more efficiently than typical demand caching schemes.

Although you can use CEF in any part of a network, it is designed for high-performance, highly resilient Layer 3 IP backbone switching. For example, Figure 4-1 shows CEF being run on Cisco 12000 series Gigabit switch routers (GSRs) at aggregation points at the core of a network where traffic levels are dense and performance is critical.

In a typical high-capacity Internet service provider environment, Cisco 12012 GSRs as aggregation devices at the core of the network support links to Cisco 7500 series routers or other feeder devices. CEF in these platforms at the network core provides the performance and scalability needed to respond to continued growth and steadily increasing network traffic. CEF is a distributed switching mechanism that scales linearly with the number of interface cards and bandwidth installed in the router.

Figure 4-1 *Cisco Express Forwarding*

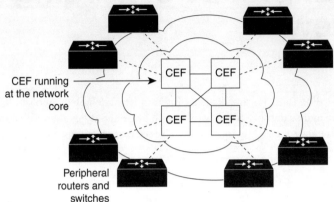

CEF Components

Information conventionally stored in a route cache is stored in several data structures for CEF switching. The data structures provide optimized lookup for efficient packet forwarding. The two main components of CEF operation are the

- Forwarding Information Base
- Adjacency Tables

Forwarding Information Base

CEF uses a Forwarding Information Base (FIB) to make IP destination prefix-based switching decisions. The FIB is conceptually similar to a routing table or an information base. It maintains a mirror image of the forwarding information contained in the IP routing table. When routing or topology changes occur in the network, the IP routing table is updated, and those changes are reflected in the FIB. The FIB maintains next-hop address information based on the information in the IP routing table.

Because there is a one-to-one correlation between FIB entries and routing table entries, the FIB contains all known routes and eliminates the need for route cache maintenance that is associated with earlier switching paths such as fast switching and optimum switching.

Adjacency Tables

Network nodes in the network are said to be *adjacent* if they can reach each other with a single hop across a link layer. In addition to the FIB, CEF uses adjacency tables to prepend Layer 2 addressing information. The adjacency table maintains Layer 2 next-hop addresses for all FIB entries.

Adjacency Discovery

The adjacency table is populated as adjacencies are discovered. Each time an adjacency entry is created (such as through the ARP protocol), a link-layer header for that adjacent node is precomputed and stored in the adjacency table. When a route is determined, it points to a next hop and corresponding adjacency entry. It is subsequently used for encapsulation during CEF switching of packets.

Adjacency Resolution

A route might have several paths to a destination prefix, such as when a router is configured for simultaneous load balancing and redundancy. For each resolved path, a pointer is added for the adjacency corresponding to the next-hop interface for that path. This mechanism is used for load balancing across several paths.

Adjacency Types That Require Special Handling

In addition to adjacencies associated with next-hop interfaces (host-route adjacencies), other types of adjacencies are used to expedite switching when certain exception conditions exist. When the prefix is defined, prefixes requiring exception processing are cached with one of the special adjacencies listed in Table 4-1.

Table 4-1 *Adjacency Types for Exception Processing*

This Adjacency Type...	Receives This Processing...
Null adjacency	Packets destined for a Null0 interface are dropped. This can be used as an effective form of access filtering.
Glean adjacency	When a router is connected directly to several hosts, the FIB table on the router maintains a prefix for the subnet rather than for the individual host prefixes. The subnet prefix point to a glean adjacency. When packets need to be forwarded to a specific host, the adjacency database is gleaned for the specific prefix.
Punt adjacency	Features that require special handling or features that are not yet supported in conjunction with CEF switching paths are forwarded to the next switching layer for handling. Features that are not supported are forwarded to the next higher switching level.
Discard adjacency	Packets are discarded. This type of adjacency occurs only on the Cisco 12000 series routers.
Drop adjacency	Packets are dropped, but the prefix is checked.

Unresolved Adjacency

When a link-layer header is suffixed to packets, FIB requires the suffix to point to an adjacency corresponding to the next hop. If an adjacency was created by FIB and not discovered through a mechanism such as ARP, the Layer 2 addressing information is not known and the adjacency is considered incomplete. Once the Layer 2 information is known, the packet is forwarded to the route processor, and the adjacency is determined through ARP.

Supported Media

CEF currently supports ATM/AAL5snap, ATM/AAL5mux, ATM/AAL5nlpid, Frame Relay, Ethernet, FDDI, PPP, HDLC, and tunnels.

CEF Operation Modes

CEF can be enabled in one of two modes:

- Central CEF Mode

- Distributed CEF Mode

Central CEF Mode

When CEF mode is enabled, the CEF FIB and adjacency tables reside on the route processor, and the route processor performs the express forwarding. You can use CEF mode when line cards are not available for CEF switching or when you need to use features not compatible with distributed CEF switching.

Figure 4-2 shows the relationship between the routing table, FIB, and adjacency table during CEF mode.

The Cisco Catalyst switches forwarding traffic from workgroup LANs to a Cisco 7500 series router on the enterprise backbone running CEF. The route processor performs the express forwarding.

Figure 4-2 *CEF Mode*

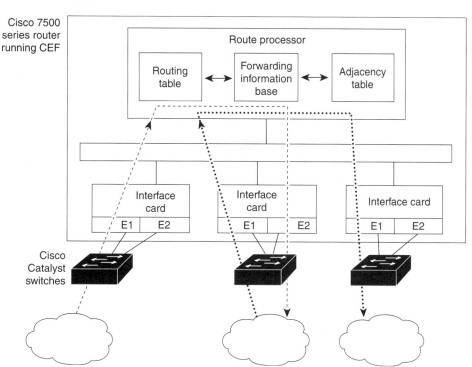

Distributed CEF Mode

When distributed CEF (dCEF) is enabled, line cards, such as VIP line cards or GSR line cards, maintain an identical copy of the FIB and adjacency tables. The line cards perform the express forwarding between port adapters, relieving the RSP of involvement in the switching operation.

dCEF uses an Inter Process Communication (IPC) mechanism to ensure synchronization of FIBs and adjacency tables on the route processor and line cards.

Figure 4-3 shows the relationship between the route processor and line cards when dCEF mode is active.

Figure 4-3 *dCEF Mode*

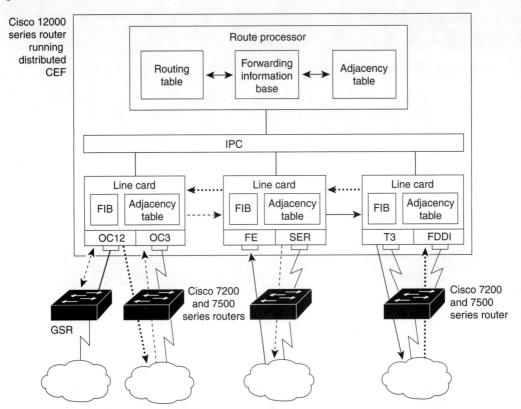

In this Cisco 12000 series router the line cards perform the switching. In other routers where you can mix various types of cards in the same router, it is possible that not all of the cards you are using support CEF. When a line card that does not support CEF receives a packet, the line card forwards the packet to the next highest switching layer (the route processor) or forwards the packet to the next hop for processing. This structure allows legacy interface processors to exist in the router with newer interface processors.

NOTE   The Cisco 12000 series GSRs operate only in dCEF mode, distributed CEF switching cannot be configured on the same VIP card as distributed fast switching, and distributed CEF is not supported on Cisco 7200 series routers.

Additional Capabilities

In addition to configuring CEF and dCEF, you can also configure these features:

- Distributed CEF switching using access lists

- Distributed CEF switching of Frame Relay packets

- Distributed CEF switching during packet fragmentation

- Load balancing on a per destination/source host pair or per packet basis

- Network accounting to gather byte and packet statistics

- Distributed CEF switching across IP tunnels

See Chapter 5, "Configuring Cisco Express Forwarding," for information on enabling these features.

Configuring Cisco Express Forwarding

This chapter describes the required and optional tasks for configuring CEF. For a complete description of CEF commands used in this chapter, refer to Chapter 6 "Cisco Express Forwarding Commands." For documentation of other commands that appear in this chapter, you can search online at www.cisco.com.

Cisco Express Forwarding Configuration Task List

The first task is required; all other tasks are optional.

- Enabling and Disabling CEF or dCEF
- Configuring Load Balancing for CEF
- Configuring Network Accounting for CEF
- Configuring Distributed Tunnel Switching for CEF

Enabling and Disabling CEF or dCEF

Enable CEF when your router has interface processors that do not support CEF.

To enable or disable CEF, use one of the following commands in global configuration mode:

Command	Purpose
ip cef	Enables standard CEF operation.
no ip cef	Disables standard CEF operation

Enable dCEF when you want your line cards to perform express forwarding so that the route processor can handle routing protocols or switch packets from legacy interface processors.

NOTE	On the Cisco 12000 series routers, dCEF is enabled by default. The command to enable dCEF is not available. Also, the configuration file does not list that dCEF is enabled on the router.

To enable or disable dCEF operation, use one of the following commands in global configuration mode:

Command	Purpose
ip cef distributed	Enables dCEF operation.
no ip cef distributed	Disables dCEF operation.

When you enable CEF or dCEF globally, all interfaces that support CEF are enabled by default. If you want to turn off CEF or dCEF on a particular interface, you can do so.

You might want to disable CEF or dCEF on a particular interface because that interface is configured with a feature that CEF or dCEF does not support. For example, policy routing and CEF cannot be used together. You might want one interface to support policy routing while the other interfaces support CEF. In this case, you would enable CEF globally, but disable CEF on the interface configured for policy routing, enabling all but one interface to express forward.

To disable CEF or dCEF on an interface, use the following command in interface configuration mode:

Command	Purpose
no ip route-cache cef	Disables CEF operation on the interface.

When you disable CEF or dCEF, Cisco IOS software switches packets by using the next fastest switching path. In the case of dCEF, the next fastest switching path is CEF on the route processor.

If you have disabled CEF or dCEF operation on an interface and want to reenable it, you can do so by using the **ip route-cache cef** command in interface configuration mode.

NOTE	On the Cisco 12000 series routers, you must not disable dCEF on an interface.

Configuring Load Balancing for CEF

CEF load balancing is based on a combination of source and destination packet information; it allows you to optimize resources by distributing traffic over multiple paths for transferring data to a destination. You can configure load balancing on a per-destination or per-packet basis. Load balancing decisions are made on the outbound interface. When you configure load balancing, configure it on outbound interfaces.

Load Balancing Configuration Task List

These sections describe how to configure each type of load balancing:

● Configuring Per-Destination Load Balancing

● Configuring Per-Packet Load Balancing

Configuring Per-Destination Load Balancing

Per-destination load balancing allows the router to use multiple paths to achieve load sharing. Packets for a given source/destination host pair are guaranteed to take the same path, even if multiple paths are available. Traffic destined for different pairs tends to take different paths. Per-destination load balancing is enabled by default when you enable CEF, and is the load balancing method of choice for most situations.

Because per-destination load balancing depends on the statistical distribution of traffic, load sharing becomes more effective as the number of source/destination pairs increases.

You can use per-destination load balancing to ensure that packets for a given host pair arrive in order. All packets for a certain host pair are routed over the same link (or links).

Enabling Per-Destination Load Balancing

Per-destination load balancing is enabled by default when you enable CEF. To use per-destination load balancing, you do not perform any additional tasks once you enable CEF.

Disabling Per-Destination Load Balancing

Typically, you would disable per-destination load balancing when you want to enable per-packet load balancing.

To disable per-destination load balancing, use the following command in interface configuration mode:

Command	Purpose
no ip load-sharing per-destination	Disables per-destination load balancing.

Configuring Per-Packet Load Balancing

Per-packet load balancing allows the router to send successive data packets over paths without regard to individual hosts or user sessions. It uses the round-robin method to determine which path each packet takes to the destination. Per-packet load balancing ensures balancing over multiple links.

Path utilization with per-packet load balancing is good, but packets for a given source/destination host pair might take different paths. Per-packet load balancing could introduce reordering of packets. This type of load balancing would be inappropriate for certain types of data traffic (such as voice traffic over IP) that depend on packets arriving at the destination in sequence.

Use per-packet load balancing to help ensure that a path for a single source/destination pair does not get overloaded. If the bulk of the data passing through parallel links is for a single pair, per-destination load balancing will overload a single link while other links have very little traffic. Enabling per-packet load balancing allows you to use alternate paths to the same busy destination.

To enable per-packet load balancing, use the following command in interface configuration mode:

Command	Purpose
ip load-sharing per-packet	Enables per-packet load balancing.

NOTE	If you want to enable per-packet load sharing to a particular destination, then all interfaces that can forward traffic to the destination must be enabled for per-packet load sharing.

Configuring Network Accounting for CEF

You might want to collect statistics to better understand CEF patterns in your network. For example, you might want to collect information such as the number of packets and bytes switched to a destination or the number of packets switched through a destination.

You can configure network accounting for CEF by performing these optional tasks:

● Enabling Network Accounting for CEF

● Viewing Network Accounting Information

Enabling Network Accounting for CEF

To collect network accounting information for CEF, use the following commands in global configuration mode:

Command	Purpose
ip cef accounting per-prefix	Enables the collection of the number of packets and bytes express forwarded to a destination (or prefix).
ip cef accounting non-recursive	Enables the collection of the number of packets express forwarded through a destination.

When you enable network accounting for CEF from global configuration mode, accounting information is collected at the route processor when CEF mode is enabled. When network accounting is enabled for dCEF, information is collected at the line cards.

Viewing Network Accounting Information

You can view the collected accounting information. To do so, use the following command in EXEC mode:

Command	Purpose
show ip cef	Displays the collected accounting information.

Configuring Distributed Tunnel Switching for CEF

CEF supports distributed tunnel switching, such as GRE tunnels. Distributed tunnel switching is enabled automatically when you enable CEF or dCEF. You do not perform any additional tasks to enable distributed tunnel switching after you enable CEF or dCEF.

CHAPTER 6

Cisco Express Forwarding Commands

This chapter documents commands used to configure CEF in Cisco IOS software. For guidelines on configuring CEF, refer to Chapter 5, "Configuring Cisco Express Forwarding."

NOTE   Beginning with Cisco IOS Release 11.3, all commands supported on the Cisco 7500 series routers are also supported on Cisco 7000 series routers.

---

clear adjacency

To clear the CEF adjacency table, use the **clear adjacency** EXEC command.

   clear adjacency

Syntax Description

This command has no arguments or keywords.

Command Mode

EXEC

Usage Guidelines

This command first appeared to support the Cisco 12012 GSR in Cisco IOS Release 11.2 GS and first appeared with multiple platform support in Cisco IOS Release 11.1 CC.

When you issue this command, entries in the adjacency table that resides on the route processor are removed and then repopulated. During repopulation, Layer 2 next hop information is reevaluated.

With dCEF mode, the adjacency tables that reside on line cards are always synchronized to the adjacency table that resides on the route processor. Therefore, clearing the adjacency table on the route processor using the **clear adjacency** command also clears the adjacency tables on the line cards; all changes are propagated to the line cards.

Example

The following example clears the adjacency table:

```
clear adjacency
```

Related Commands

You can search online at www.cisco.com to find documentation of related commands.

show adjacency

clear cef linecard

To clear CEF information from line cards, use the **clear cef linecard** EXEC command.

clear cef linecard [*slot-number*] [**adjacency** | **interface** | **prefix**]

Syntax	Description
slot-number	(Optional) Line card slot number to clear. When you omit this argument, all line card slots are cleared.
adjacency	(Optional) Clears line card adjacency tables and rebuilds adjacency for the specified line card.
interface	(Optional) Clears line card interface information and re-creates the interface information for the specified line card.
prefix	(Optional) Clears line card prefix tables and starts rebuilding the FIB table.

Command Mode

EXEC

Usage Guidelines

This command first appeared to support the Cisco 12012 GSR in Cisco IOS Release 11.2 GS and first appeared with multiple platform support in Cisco IOS Release 11.1 CC.

This command is available only on routers with line cards. This command clears CEF information only on the line cards; CEF information on the route processor is not affected.

After you clear CEF information from line cards, the corresponding information from the route processor is propagated to the line cards. IPC ensures that CEF information on the route processor matches the CEF information on the line cards.

Example

The following example clears the CEF information from the line cards:

```
clear cef linecard
```

Related Commands

You can search online at www.cisco.com to find documentation of related commands.

show cef linecard

clear ip cef prefix-statistics

To clear CEF counters by resetting the packet and byte count to zero (0), use the **clear ip cef prefix-statistics** EXEC command.

clear ip cef {*network* [*mask*] | *** } **prefix-statistics**

Syntax	Description
network	Clears counters for a FIB entry specified by network.
mask	(Optional) Clears counters for a FIB entry specified by network and mask.
*	Clears counters for all FIB entries.

Command Mode

EXEC

Usage Guidelines

This command first appeared to support the Cisco 12012 GSR in Cisco IOS Release 11.2 GS and first appeared with multiple platform support in Cisco IOS Release 11.1 CC.

Example

The following example resets the CEF packet and byte count to zero:

```
clear ip cef prefix-statistics
```

Related Commands

You can search online at www.cisco.com to find documentation of related commands.

show adjacency
show ip cef

ip cef

To enable CEF on the route processor card, use the **ip cef** global configuration command. To disable CEF, use the **no** form of this command.

> **ip cef** [**distributed**]
> **no ip cef** [**distributed**]

Syntax

Description

distributed

(Optional) Enables dCEF operation. Distributes CEF information to line cards. Line cards perform express forwarding.

Defaults

On This Platform...	The Default Is...
Cisco 7000 series equipped with RSP7000	CEF is not enabled.
Cisco 7200 series	CEF is not enabled.
Cisco 7500 series	CEF is enabled.
Cisco 12000 series GSR	dCEF is enabled.

Command Mode

Global configuration

Usage Guidelines

This command first appeared in Cisco IOS Release 11.1 CC.

This command is not available on the Cisco 12000 series GSR because that router series operates only in dCEF mode.

CEF is advanced Layer 3 IP switching technology. CEF optimizes network performance and scalability for networks with dynamic, topologically dispersed traffic patterns, such as those associated with Web-based applications and interactive sessions.

Examples

The following example enables standard CEF operation:

```
ip cef
```

The following example enables dCEF operation:

```
ip cef distributed
```

Related Commands

You can search online at www.cisco.com to find documentation of related commands.

ip route-cache cef

ip cef accounting

To enable network accounting of CEF, use the **ip cef accounting** global configuration command. To disable network accounting of CEF, use the **no** form of this command.

> **ip cef accounting [per-prefix] [non-recursive]**
> **no ip cef accounting [per-prefix] [non-recursive]**

Syntax	Description
per-prefix	(Optional) Enables the collection of the number of packets and bytes express forwarded to a destination (or prefix).
non-recursive	(Optional) Enables accounting through non-recursive prefixes. For prefixes with directly connected next hops, enables the collection of the number of packets and bytes express forwarded through a prefix.

Default

Accounting is disabled by default.

Command Mode

Global configuration

Usage Guidelines

This command first appeared to support the Cisco 12012 GSR in Cisco IOS Release 11.2 GS, and first appeared with multiple platform support in Cisco IOS Release 11.1 CC.

You might want to collect statistics to better understand CEF patterns in your network.

When you enable network accounting for CEF from global configuration mode, accounting information is collected at the route processor when CEF mode is enabled and at the line cards when dCEF mode is enabled.

You can then view the collected accounting information using the **show ip cef** command.

Example

The following example enables the collection of CEF accounting information:

```
ip cef accounting
```

Related Commands

You can search online at www.cisco.com to find documentation of related commands.

show ip cef

ip cef traffic-statistics

To change the time interval that controls when NHRP will set up or tear down an SVC, use the **ip cef traffic-statistics** global configuration command. To restore the default values, use the **no** form of this command.

> **ip cef traffic-statistics** [**load-interval** *seconds*] [**update-rate** *seconds*]
> **no ip cef traffic-statistics**

Syntax	Description
load-interval *seconds*	(Optional) Length of time (in 30-second increments) during which the average *trigger-threshold* and *teardown-threshold* are calculated before an SVC setup or teardown action is taken. (These thresholds are configured in the **ip nhrp trigger-svc** command.) The **load-interval** range is 30 seconds to 300 seconds, in 30-second increments.
update-rate *seconds*	(Optional) Frequency that the port adapter sends the accounting statistics to the RP. When using NHRP in distributed CEF switching mode, this value must be set to 5 seconds.

Default

load-interval: 30 seconds
update-rate: 10 seconds

Command Mode

Global configuration

Usage Guidelines

This command first appeared in Cisco IOS Release 12.0.

The thresholds in the **ip nhrp trigger-svc** command must be exceeded during a certain time interval, which is 30 seconds by default. To change that interval, use the **load-interval** *seconds* argument of the **ip cef traffic-statistics** command.

When NHRP is configured on a CEF switching node with a VIP2 adapter, you must make sure the **update-rate** is set to 5 seconds.

Other features could also use the **ip cef traffic-statistics** command; this NHRP feature relies on it.

Example

In the following example, the triggering and teardown thresholds are calculated based on an average over 120 seconds:

```
ip cef traffic-statistics load-interval 120
```

Related Commands

You can search online at www.cisco.com to find documentation of related commands.

ip nhrp trigger-svc

ip load-sharing

To enable load balancing for CEF, use the **ip load-sharing** interface configuration command. To disable load balancing for CEF, use the **no** form of this command.

> **ip load-sharing [per-packet] [per-destination]**
> **no ip cef [per-packet]**

Syntax	Description
per-packet	(Optional) Enables per-packet load balancing on the interface.
per-destination	(Optional) Enables per-destination load balancing on the interface.

Default

Per-destination load balancing is enabled by default when you enable CEF.

Command Mode

Interface configuration

Usage Guidelines

This command first appeared to support the Cisco 12012 GSR in Cisco IOS Release 11.2 GS, and first appeared with multiple platform support in Cisco IOS Release 11.1 CC.

Per-packet load balancing allows the router to send data packets over successive equal-cost paths without regard to individual destination hosts or user sessions. Path utilization is good, but packets destined for a given destination host might take different paths and might arrive out of order.

Per-destination load balancing allows the router to use multiple, equal-cost paths to achieve load sharing. Packets for a given source/destination host pair are guaranteed to take the same path, even if multiple, equal-cost paths are available. Traffic for different source/destination host pairs tends to take different paths.

NOTE   If you want to enable per-packet load sharing to a particular destination, then all interfaces that can forward traffic to the destination must be enabled for per-packet load sharing.

Examples

The following example enables per-packet load balancing:

```
interface E0
 ip load-sharing per-packet
```

The following example enables per-destination load balancing:

```
interface E0
 ip load-sharing per-destination
```

Related Commands

You can search online at www.cisco.com to find documentation of related commands.

interface
ip cef

ip route-cache cef

To enable CEF operation on an interface after CEF operation has been disabled, use the **ip route-cache cef** interface configuration command. To disable CEF operation on an interface, use the **no** form of this command.

> **ip route-cache cef**
> **no ip route-cache cef**

Syntax Description

This command has no arguments or keywords.

Defaults

When standard CEF or dCEF operation is enabled globally, all interfaces that support CEF are enabled by default.

Command Mode

Interface configuration

Usage Guidelines

This command first appeared to support the Cisco 12012 GSR in Cisco IOS Release 11.2 GS, and first appeared with multiple platform support in Cisco IOS Release 11.1 CC.

CEF is an advanced Layer 3 switching technology for IP. CEF optimizes network performance and scalability for networks with dynamic, topologically dispersed traffic patterns, such as those associated with Web-based applications and interactive type sessions.

Because all interfaces that support CEF or dCEF are enabled by default when you enable standard CEF or dCEF operation globally, you use the **no** form of the command to turn off CEF operation on a particular interface.

You might want to disable CEF or dCEF on a particular interface because that interface is configured with a feature that CEF or dCEF does not support. For example, policy routing and CEF cannot be used together. You might want one interface to support policy routing while the other interfaces support CEF. In this case, you would turn on CEF globally, but turn off CEF on the interface configured for policy routing, enabling all but one interface to express forward.

When you disable CEF or dCEF, Cisco IOS software switches packets using the next fastest switching path. In the case of dCEF, the next fastest switching path is CEF on the route processor.

If you have disabled CEF or dCEF operation on an interface and want to reenable it, you can do so by using the **ip route-cache cef** command in interface configuration mode.

NOTE On the Cisco 12000 series routers, you must not disable dCEF on an interface.

Examples

The following example enables CEF operation on the router (globally), but turns off CEF operation on Ethernet interface 0:

```
ip cef
interface e0
 no ip route-cache cef
```

The following example enables dCEF operation on the router (globally), but turns off CEF operation on Ethernet interface 0:

```
ip cef distributed
interface e0
 no ip route-cache cef
```

The following example reenables dCEF operation on Ethernet interface 0:

```
ip cef distributed
interface e0
 ip route-cache cef
```

Related Commands

You can search online at www.cisco.com to find documentation of related commands.

interface
ip cef

show adjacency

To display CEF adjacency table information, use the **show adjacency** EXEC command.

       show adjacency [**detail**]

Syntax	Description
detail	(Optional) Displays detailed adjacency information, including Layer 2 information.

Command Mode

EXEC

Usage Guidelines

This command first appeared to support the Cisco 12012 GSR in Cisco IOS Release 11.2 GS, and first appeared with multiple platform support in Cisco IOS Release 11.1 CC.

This command is available only on routers that have RP cards.

Sample Display

The following is sample output from the **show adjacency detail** command:

```
Router# show adjacency detail

Protocol   Interface            Address
IP         Tunnel0              point2point(3)   (incomplete)
                                0 packets, 0 bytes
                                FIB        00:02:45
IP         Ethernet1/0/0        192.168.177.15(6)
                                0 packets, 0 bytes
                                0060837BEFA0

Protocol   Interface            Address
                                0060836FA7000800
                                ARP        03:59:44
                                igrp 622   00:04:14
IP         Ethernet0/0          192.168.233.88(5)
                                0 packets, 0 bytes
                                0060837BEFA0
                                0060836FA7000800
                                ARP        03:59:36
IP         FastEthernet2/0/0    172.16.1.106 (11)   (incomplete)
                                0 packets, 0 bytes
IP         FastEthernet2/0/0    172.26.1.106 (11)   (incomplete)
                                0 packets, 0 bytes
```

Table 6-1 describes the fields shown in the output from the **show adjacency detail** command.

Table 6-1 *show adjacency detail Command Field Descriptions*

Field	Description
Protocol	The routing protocol configured on the interface.
Interface	The type of interface configured.
Address	The address of the interface.
Routing protocol	The method by which the adjacency was learned.
Adjacent next hop	The MAC address of the adjacent router.
Time stamp	The time left before the adjacency rolls out of the adjacency table. After it rolls out, a packet must use the same next hop to the destination.

Related Commands

You can search online at www.cisco.com to find documentation of related commands.

clear adjacency

show cef

To display which packets the line cards dropped or to display which packets were not express forwarded, use the **show cef** EXEC command.

show cef [drop | not-cef-switched]

Syntax	Description
drop	(Optional) Displays which packets were dropped by each line card.
not-cef-switched	(Optional) Displays which packets were sent to a different switching path.

Command Mode

EXEC

Usage Guidelines

This command first appeared to support the Cisco 12012 GSR in Cisco IOS Release 11.2 GS, and first appeared with multiple platform support in Cisco IOS Release 11.1 CC.

This command is available only on routers that have RP cards.

A line card might drop packets due to encapsulation failure, no route information, or no adjacency information.

A packet is sent to a different switching path because CEF does not support the encapsulation or feature, the packet is destined for the router, or the packet has IP options, such as time stamp and record route. IP options are process switched.

Sample Displays

The following is sample output from the **show cef drop** command:

```
Router# show cef drop

CEF Drop Statistics
Slot  Encap_fail  Unresolved Unsupported   No_route    No_adj   ChksumErr
RP             4          89           0          4         0           0
1              0           0           0          0         0           0
2              0           0           5          0         0           5
```

Table 6-2 describes the fields shown in the output from the **show cef drop** command.

Table 6-2 *show cef drop Field Descriptions*

Field	Description
Slot	The slot number on which the packets were received.
Encap_fail	Indicates the number of packets dropped after the limit was reached for incomplete packets with no adjacency route.
Unresolved	Indicates the number of packets dropped because the route for the prefix was not resolved.
Unsupported	Indicates the number of packets received for which the adjacency route information was dropped due to unsupported features.
No_route	No route definition is included in the prefix table.
No_adj	The prefix is resolved, but the adjacent route is not indicated.
ChksumErr	Indicates the number of packets received with a checksum error.

The following is sample output from the **show cef not-cef-switched** command:

```
Router# show cef not-cef-switched

CEF Packets passed on to next switching layer
Slot No_adj No_encap Unsuppted Redirect Receive Bad_ttl Options Access RP
0         0        0         0        0   91584       0       0      0   0
1         0        0         0        0       0       0       0      0   0    0
2         0        0         0        0       0       0       0      0   0    0
```

Table 6-3 describes the fields shown in the output from the **show cef not-cef-switched** command.

Table 6-3 *show cef not-cef-switched Field Descriptions*

Field	Meaning
No_adj	Indicates the number of packets sent to the line card to ARP for the adjacent route.
No_encap	Number of encapsulated packets received.
Unsupported Redirect	Number of packets with unsupported features and redirected to another switching layer or location for processing.

Related Commands

You can search online at www.cisco.com to find documentation of related commands.

show cef interface
show cef linecard

show cef interface

To display CEF-related interface information, use the **show cef interface** EXEC command.

> **show cef interface** *type number* [**detail**]

Syntax	Description
type number	Interface type and number about which to display CEF-related information.
detail	(Optional) Displays detailed CEF information for the specified interface type and number.

Command Mode

EXEC

Usage Guidelines

This command first appeared to support the Cisco 12012 GSR in Cisco IOS Release 11.2 GS, and first appeared with multiple platform support in Cisco IOS Release 11.1 CC.

This command is available on routers that have RP cards and line cards.

The **detail** command displays more CEF-related information for the specified interface.

You can use this command to show the CEF state on an individual interface.

Sample Displays

The following is sample output from the **show cef interface detail** command for Ethernet interface 0:

```
Router# show cef interface E0 detail

Ethernet1/0/0 is up (if_number 6)
Internet address is 172.19.177.20/24
ICMP redirects are always sent
Per-packet load balancing is disabled
Inbound access list is 10
Outbound access list is not set
Hardware idb is Ethernet1/0/0
Fast switching type 1, interface type 5
IP Distributed CEF switching enabled
IP Feature CEF switching turbo vector
Fast flags 0x4. ifindex 5(5)
Slot 1 Slot unit 0 VC -1
Hardware transmit queue ptr 0x48001A00 (0x48001A00) >- debugging purposes Transmit limit
accumulator 0x48001A02 (0x48001A02) IP MTU 1500
```

Table 6-4 describes the fields shown in the output from the **show cef interface detail** command for Ethernet interface 0.

Table 6-4 *show cef interface detail Field Descriptions*

Field	Description
interface type number is {up | down}	Indicates status of the interface.
Internet address	Internet address of the interface.
ICMP packets are {always sent | never sent}	Indicates how packet forwarding is configured.
Per-packet load balancing	Status of load balancing in use on the interface (enabled or disabled).
Inbound access list {# | Not set}	Number of access lists defined for the interface.
Outbound access list	Number of access lists defined for the interface.
Hardware idb is *type number*	Interface type and number configured.
Fast switching type	Used for troubleshooting; indicates the switching mode in use.
IP Distributed CEF switching {enabled | disabled}	Indicates the switching path used.
Slot *n* Slot unit *n*	The slot number.
Hardware transmit queue	Indicates the number of packets in the transmit queue.
Transmit limit accumulator	Indicates the maximum number of packets allowed in the transmit queue.
IP MTU	The value of the MTU size set on the interface.

Related Commands

You can search online at www.cisco.com to find documentation of related commands.

show cef
show cef linecard

show cef linecard

To display CEF-related interface information by line card, use the **show cef linecard** EXEC command.

show cef linecard [*slot-number*] [**detail**]

Syntax

Description

slot-number

(Optional) Slot number containing the line card about which to display CEF-related information. When you omit this argument, information about all line cards is displayed.

detail

(Optional) Displays detailed CEF information for the specified line card.

Command Mode

EXEC

Usage Guidelines

This command first appeared to support the Cisco 12012 GSR in Cisco IOS Release 11.2 GS, and first appeared with multiple platform support in Cisco IOS Release 11.1 CC.

This command is available only on routers that have RP cards.

When you omit the *slot-number* argument, information about all line cards is displayed. When you omit the *slot-number* argument and include the **detail** keyword, detailed information is displayed for all linecards. When you omit all keywords and arguments, the **show cef linecard** command displays important information about all line cards in table format.

Sample Displays

The following is sample output from the **show cef linecard** command. The command displays information for the line cards.

```
Router# show cef linecard

CEF table version 115705, 45877 routes
Slot CEF-ver MsgSent XdrSent Seq MaxSeq LowQ HighQ Flags
```

```
1       238     668     9641 616     616     0       0 up, sync
2       238     683     10782 619    629     0       0 up, sync
```

Table 6-5 describes the fields shown in the output from the **show cef linecard** command.

Table 6-5 *show cef linecard* Field Descriptions

Field	Description
CEF table version	The FIB table version.
XdrSent	IPC information elements (xdrs) packed into IPC messages sent from the RP to the line card.
MsgSent	Number of IPC messages sent.
Seq	Sequence number for the line card.
MaxSeq	Maximum sequence expected by the line card.
LowQ/HighQ	Number of xdr elements in LowQ and HighQ.
Flags	Indicates the status of the line card. Possible states are
	• up Line card is up.
	• sync Line card is in sync with main FIB.
	• repopulate Repopulate FIB on line card.
	• reset Line card FIB is reset.
	• reloading Line card FIB is currently being reloaded.
	• disabled Line card is disabled.

Part II

Command Reference

The following is sample output from the **show cef linecard detail** command for the line card in slot number 2:

```
Router# show cef linecard 2 detail

CEF line card slot number 2, status up, sync, disabled
line card CEF version number 238
Sequence number 616, Maximum sequence number expected 616
Send failed 0, Out Of Sequence 0
line card CEF reset 2, reloaded 2
92299/15/91 prefix/adjacency/interface elements queued
49641 elements packed in 668 messages(1341286 bytes) sent
0/0 xdr elements in LowQ/HighQ
Input packets 0, bytes 0<--- line card stats
Output packets 0, bytes 0, drops 0
```

Related Commands

You can search online at www.cisco.com to find documentation of related commands.

show cef
show cef interface

show ip cef

To display entries in the FIB that are unresolved or to display a summary of the FIB, use this form of the **show ip cef** EXEC command:

show ip cef [**unresolved** | **summary**]

To display specific entries in the FIB based on IP address information, use this form of the **show ip cef** EXEC command:

show ip cef [*network* [*mask* [**longer-prefix**]]] [**detail**]

To display specific entries in the FIB based on interface information, use this form of the **show ip cef** EXEC command:

show ip cef [*type number*] [**detail**]

Syntax	Description
unresolved	(Optional) Displays unresolved FIB entries.
summary	(Optional) Displays a summary of the FIB.
network	(Optional) Displays the FIB entry for the specified destination network.
mask	(Optional) Displays the FIB entry for the specified destination network and mask.
longer-prefix	(Optional) Displays FIB entries for all more specific destinations.
detail	(Optional) Displays detailed FIB entry information.
type number	(Optional) Interface type and number for which to display FIB entries.

Command Mode

EXEC

Usage Guidelines

This command first appeared to support the Cisco 12012 GSR in Cisco IOS Release 11.2 GS, and first appeared with multiple platform support in Cisco IOS Release 11.1 CC.

The **show ip cef** command without any keywords or arguments shows a brief display of all FIB entries.

The **show ip cef detail** command shows detailed FIB entry information for all FIB entries.

Sample Displays

The following is sample output from the **show ip cef unresolved** command:

```
Router# show ip cef unresolved

IP Distributed CEF with switching (Table Version 136632)
45776 routes, 13 unresolved routes (0 old, 13 new)
45776 leaves, 2868 nodes, 8441480 bytes, 136632 inserts, 90856 invalidations
1 load sharing elements, 208 bytes, 1 references
1 CEF resets, 1 revisions of existing leaves
refcounts: 527292 leaf, 465617 node

148.214.0.0/16, version 136622
0 packets, 0 bytes
  via 171.69.233.56, 0 dependencies, recursive
  unresolved
148.215.0.0/16, version 136623
0 packets, 0 bytes
  via 171.69.233.56, 0 dependencies, recursive
  unresolved
148.218.0.0/16, version 136624
0 packets, 0 bytes
```

The following is sample output from the **show ip cef summary** command:

```
Router# show ip cef summary

IP Distributed CEF with switching (Table Version 135165)
45788 routes, 0 reresolve, 4 unresolved routes (0 old, 4 new)
45788 leaves, 2868 nodes, 8442864 bytes, 135165 inserts, 89377 invalidations
0 load sharing elements, 0 bytes, 0 references
1 CEF resets, 0 revisions of existing leaves
refcounts: 527870 leaf, 466167 node
```

The following is sample output from the **show ip cef internal** command; it shows load-sharing details for multiple paths to a prefix:

```
Router# show ip cef 192.168.1.0 internal
192.168.1.0/24, version 135490, per-destination sharing 0 packets, 0 bytes

via 172.19.233.50, 0 dependencies, recursive<-- possible path 1 info
traffic share 1, current path
next hop 172.19.233.50, Ethernet0/0 via 172.19.233.50/32 valid adjacency
via 172.19.233.49, 0 dependencies, recursive<-- possible path 2 info
traffic share 1
next hop 172.19.233.49, Ethernet0/0 via 172.19.233.49/32 valid adjacency
```

```
0 packets, 0 bytes switched through the prefix Load distribution: 0 1 0 1 0 1 0 1 0 1 0 1 0
1 0 1 (refcount 1) ^
I.. how the load is distributed among the possible paths

Hash OK InterfaceAddressPackets
1 Y Ethernet0/0172.19.233.500
2 Y Ethernet0/0172.19.233.490
3 Y Ethernet0/0172.19.233.500
4 Y Ethernet0/0172.19.233.490
5 Y Ethernet0/0172.19.233.500
6 Y Ethernet0/0172.19.233.490
7 Y Ethernet0/0172.19.233.500
8 Y Ethernet0/0172.19.233.490
9 Y Ethernet0/0172.19.233.500
10 Y Ethernet0/0172.19.233.490
11 Y Ethernet0/0172.19.233.500
12 Y Ethernet0/0172.19.233.490
13 Y Ethernet0/0172.19.233.500
14 Y Ethernet0/0172.19.233.490
15 Y Ethernet0/0172.19.233.500
16 Y Ethernet0/0172.19.233.490
```

The following is sample output from the **show ip cef detail** command for Ethernet interface 0. It shows all the prefixes resolving through adjacency pointing to next hop Ethernet interface 0/0 and next-hop interface IP address 172.19.233.33.

```
Router# show ip cef e0/0 172.19.233.33 detail
IP Distributed CEF with switching (Table Version 136808)
45800 routes, 8 unresolved routes (0 old, 8 new) 45800 leaves, 2868 nodes, 8444360 bytes,
136808 inserts, 91008 invalidations 1 load sharing elements, 208 bytes, 1 references 1 CEF
resets, 1 revisions of existing leaves refcounts: 527343 leaf, 465638 node

172.19.233.33/32, version 7417, cached adjacency 172.19.233.33 0 packets, 0 bytes, Adjacency-
prefix
via 172.19.233.33, Ethernet0/0, 0 dependencies
next hop 172.19.233.33, Ethernet0/0
valid cached adjacency
```

Related Commands

You can search online at www.cisco.com to find documentation of related commands.

show cef
show cef interface

PART III

NetFlow Switching

CHAPTER 7

NetFlow Switching Overview

This chapter describes NetFlow switching.

NetFlow Switching

NetFlow switching provides network administrators with access to "call detail recording" information from their data networks. Exported NetFlow data can be used for a variety of purposes, including network management and planning, enterprise accounting and departmental chargebacks, ISP billing, data warehousing, and data mining for marketing purposes. NetFlow also provides a highly efficient mechanism with which to process security access lists without paying as much of a performance penalty as is incurred with other available switching methods.

This chapter describes NetFlow switching. It contains these sections:

- NetFlow Switching Support
- Accounting Statistics
- The NetFlow Data Format

NetFlow Switching Support

NetFlow switching is supported on Cisco 7200 series routers and Cisco 7500 series routers.

Accounting Statistics

NetFlow switching is a high-performance, network-layer switching path that captures as part of its switching function a rich set of traffic statistics. These traffic statistics include user, protocol, port, and type of service information that can be used for a wide variety of purposes such as network analysis and planning, accounting, and billing.

NetFlow switching is supported on IP and IP-encapsulated traffic over all interface types and encapsulations except for ISL/VLAN, ATM, and Frame Relay interfaces when more than one input access control list is used on the interface, and ATM LANE.

Capturing Traffic Data

In conventional switching at the network layer, each incoming packet is handled on an individual basis with a series of functions to perform access list checks, capture accounting data, and switch the packet. With NetFlow switching, after a flow has been identified and access list processing of the first packet in the flow has been performed, all subsequent packets are handled on a connection-oriented basis as part of the flow, where access list checks are bypassed and packet switching and statistics capture are performed in tandem.

A network flow is identified as a unidirectional stream of packets between a given source and destination—both defined by a network-layer IP address and transport-layer port number. Specifically, a flow is identified as the combination of the following fields:

- Source IP address

- Destination IP address

- Source port number

- Destination port number

- Protocol type

- Type of service

- Input interface

The NetFlow Cache

NetFlow switching operates by creating a flow cache that contains the information needed to switch and perform access list checks for all active flows. The NetFlow cache is built by processing the first packet of a flow through the standard switching path. As a result, each flow is associated with an incoming and outgoing interface port number and with a specific security access permission and encryption policy. The cache also includes entries for traffic statistics that are updated in tandem with the switching of subsequent packets. After the NetFlow cache is created, packets identified as belonging to an existing flow can be switched, based on the cached information and security access list checks bypassed. Flow information is maintained within the NetFlow cache for all active flows.

The NetFlow Data Format

NetFlow exports flow information in UDP datagrams in one of two formats. The version 1 format was the initial release version, and the version 5 format was a later enhancement to add Border Gateway Protocol (BGP) AS information and flow sequence numbers. Versions 2 through 4 were not released.

In version 1 and version 5 format, the datagram consists of a header and one or more flow records. The first field of the header contains the version number of the export datagram. Typically a receiving application that accepts either format allocates a buffer large enough for the biggest possible datagram from either format and uses the version from the header to determine how to interpret the datagram. The second field in the header is the number of records in the datagram and should be used to index through the records.

All fields in either version 1 or version 5 format are in network byte order. Table Tables 7-1 and 7-2 describe the data format for version 1, and Tables 7-3 and 7-4 describe the data format for version 5.

Cisco recommends that receiving applications check datagrams to ensure that the datagrams are from a valid NetFlow source. Cisco recommends that you first check the size of the datagram to make sure it is at least long enough to contain the version and count fields. Next, you should verify that the version is valid (1 or 5) and that the number of received bytes is enough for the header and count flow records (using the appropriate version).

Because NetFlow export uses User Datagram Protocol (UDP) to send export datagrams, it is possible for datagrams to be lost. To determine whether flow export information is lost, the version 5 header format contains a flow sequence number. The sequence number is equal to the sequence number of the previous plus the number of flows in the previous datagram. After receiving a new datagram, the receiving application can subtract the expected sequence number from the sequence number in the header to get the number of missed flows. Table 7-1 lists the bytes for version 1 header format.

Table 7-1   *Version 1 Header Format*

Bytes	Content	Description
0-3	version and count	Netflow export format version number and number of flows exported in this packet (1-24).
4-7	SysUptime	Current time in milliseconds since router booted.
8-11	unix_secs	Current seconds since 0000 UTC 1970.
12-16	unix_nsecs	Residual nanoseconds since 0000 UTC 1970.

Table 7-2 lists the byte definitions for version 1 flow record format.

Table 7-2   *Version 1 Flow Record Format*

Bytes	Content	Description
0-3	srcaddr	Source IP address.
4-7	dstaddr	Destination IP address.
8-11	nexthop	Next hop router's IP address.
12-15	input and output	Input and output interface's SNMP index.
16-19	dPkts	Packets in the flow.
20-23	dOctets	Total number of Layer 3 bytes in the flow's packets.
24-27	First	SysUptime at start of flow.
28-31	Last	SysUptime at the time the last packet of flow was received.
32-35	srcport and dstport	TCP/UDP source and destination port number or equivalent.
36-39	pad1, prot, and tos	Unused (zero) byte, IP protocol (for example, 6=TCP, 17=UDP), and IP type-of-service.
40-43	flags, pad2, and pad3	Cumulative OR of TCP flags. Pad 2 and pad 3 are unused (zero) byte.

Table 7-2 *Version 1 Flow Record Format*

Bytes	Content	Description
44-48	reserved	Unused (zero) bytes.

Table 7-3 lists the byte definitions for version 5 header format.

Table 7-3 *Version 5 Header Format*

Bytes	Content	Description
0-3	version and count	Netflow export format version number and number of flows exported in this packet (1-30).
4-7	SysUptime	Current time in milliseconds since router booted
8-11	unix_secs	Current seconds since 0000 UTC 1970.
12-15	unix_nsecs	Residual nanoseconds since 0000 UTC 1970.
16-19	flow_sequence	Sequence counter of total flows seen.
20-24	reserved	Unused (zero) bytes.

Table 7-4 lists the byte definitions for version 5 flow record format.

Table 7-4 *Version 5 Flow Record Format*

Bytes	Content	Description
0-3	srcaddr	Source IP address.
4-7	dstaddr	Destination IP address.
8-11	nexthop	Next hop router's IP address.
12-15	input and output	Input and output interface's SNMP index.
16-19	dPkts	Packets in the flow.
20-23	dOctets	Total number of Layer 3 bytes in the flow's packets.
24-27	First	SysUptime at start of flow.
28-31	Last	SysUptime at the time the last packet of flow was received.
32-35	srcport and dstport	TCP/UDP source and destination port number or equivalent.
36-39	pad1, tcp_flags, prot, and tos	Unused (zero) byte, Cumulative OR of TCP flags, IP protocol (for example, 6=TCP, 17=UDP), and IP type-of-service.
40-43	src_as and dst_as	AS of the source and destination, either origin or peer.
44-48	src_mask, dst_mask, and pad2	Source and destination address prefix mask bits, pad 2 is unused (zero) bytes.

CHAPTER 8

Configuring NetFlow Switching

This chapter describes how to configure NetFlow switching. For a complete description of NetFlow commands used in this chapter, refer to Chapter 3,"Cisco IOS Switching Commands." For documentation of other commands that appear in this chapter, you can search online at www.cisco.com. This chapter contains these sections:

- Configuring NetFlow Switching
- NetFlow Switching Configuration Example

Configuring NetFlow Switching

NetFlow switching is one of the available switching modes. When you configure NetFlow on an interface, the other switching modes are not used on that interface. Also, with NetFlow switching you can export data (traffic statistics) to a remote workstation for further processing.

NetFlow switching is based on identifying packet flows and performing switching and access list processing within a router. It does not involve any connection-setup protocol either between routers or to any other networking device or end station and does not require any change externally—either to the traffic or packets themselves or to any other networking device. Thus, NetFlow switching is completely transparent to the existing network, including end stations, application software, and network devices such as LAN switches. Also, because NetFlow switching is performed independently on each internetworking device, it does not need to be operational on each router in the network. Network planners can selectively invoke NetFlow switching (and NetFlow data export) on a router or interface basis to gain traffic performance, control, or accounting benefits in specific network locations.

NOTE   NetFlow consumes more memory and CPU resources than other switching modes; therefore, it is important to understand the resources required on your router before enabling NetFlow.

To configure NetFlow switching, first configure the router for IP routing. After you configure IP routing, use the following commands, beginning in global configuration mode:

Step	Command	Purpose
1	**interface** *type slot/port-adapter/port* (Cisco 7500 series routers) or **interface** *type slot/port* (Cisco 7200 series routers)	Specifies the interface, and enters interface configuration mode.
2	**ip route-cache flow**	Specifies flow switching.

NetFlow switching information can also be exported to network management applications. To configure the router to export NetFlow switching statistics maintained in the NetFlow cache to a workstation when a flow expires, use one of the following commands in global configuration mode:

Command	Purpose
ip flow-export *ip-address udp-port* [**version 1**]	Configures the router to export NetFlow cache entries to a workstation if you are using receiving software that requires version 1. Version 1 is the default.
ip flow-export *ip-address udp-port* **version 5** [**origin-as** | **peer-as**]	Configures the router to export NetFlow cache entries to a workstation if you are using receiving software that accepts version 5. Optionally specifies origin or peer AS. The default is to export neither AS, which provides improved performance.

Normally the size of the NetFlow cache will meet your needs. However, you can increase or decrease the number of entries maintained in the cache to meet the needs of your NetFlow traffic rates. The default is 64K flow cache entries. Each cache entry is approximately 64 bytes of storage. Assuming a cache with the default number of entries, approximately 4 MB of DRAM would be required. Each time a new flow is taken from the free-flow queue, the number of free flows is checked. If there are only a few free flows remaining, NetFlow attempts to age 30 flows using an accelerated timeout. If there is only one free flow remaining, NetFlow automatically ages 30 flows regardless of their age. The intent is to ensure that free flow entries are always available.

To customize the number of entries in the NetFlow cache, use the following command in global configuration mode:

Command	Purpose
ip flow-cache entries *number*	Changes the number of entries maintained in the NetFlow cache. The number of entries can be 1024 to 524288. The default is 65536.

CAUTION	Cisco recommends that you not change the NetFlow cache entries. Improper use of this feature could cause network problems. To return to the default NetFlow cache entries, use the **no ip flow-cache entries** global configuration command.

Managing NetFlow Switching Statistics

You can display and clear NetFlow switching statistics. NetFlow statistics consist of IP packet size distribution, IP flow switching cache information, and flow information such as the protocol, total flow, flows per second, and so forth. The resulting information can be used to find out information about your router traffic. To manage NetFlow switching statistics, use either of the following commands in privileged EXEC mode:

Command	Purpose
show ip route flow	Displays the NetFlow switching statistics.
clear ip flow stats	Clears the NetFlow switching statistics.

Configuring IP Distributed and NetFlow Switching on VIP Interfaces

On Cisco 7500 series routers with an RSP and with VIP controllers, the VIP hardware can be configured to switch packets received by the VIP with no per-packet intervention on the part of the RSP. This process is called *distributed switching*. Distributed switching decreases the demand on the RSP.

The VIP hardware can also be configured for NetFlow switching, a high-performance feature that identifies initiation of traffic flow between Internet endpoints, caches information about the flow, and uses this cache for high-speed switching of subsequent packets within the identified stream.

NetFlow switching data can also be exported to network management applications.

To configure distributed switching on the VIP, first configure the router for IP routing as described in this chapter and the various routing protocol chapters, depending on the protocols you use.

After you configure IP routing, use the following commands, beginning in global configuration mode:

Step	Command	Purpose
1	**interface** *type slot/port-adapter/port*	Specifies the interface, and enter interface configuration mode.
2	**ip route-cache distributed**	Enables VIP distributed switching of IP packets on the interface.
3	**ip route-cache flow**	Specifies flow switching.

When the RSP or VIP is flow switching, it uses a flow cache instead of a destination network cache to switch IP packets. The flow cache uses source and destination network address, protocol, and source and destination port numbers to distinguish entries.

To export NetFlow switching cache entries to a workstation when a flow expires, use the following command in global configuration mode:

Command	Purpose
ip flow-export *ip-address udp-port*	Configures the router to export NetFlow cache entries to a workstation.

To improve performance, fragmented IP packets are flow switched rather than being process switched by default on Cisco 7500 series routers.

NetFlow Switching Configuration Example

The following example shows how to modify the configuration of serial interface 3/0/0 to enable NetFlow switching and to export the flow statistics for further processing to UDP port 0 on a workstation with the IP address of 1.1.15.1. In this example, existing NetFlow statistics are cleared to ensure accurate information when the **show ip cache flow** command is executed to view a summary of the NetFlow switching statistics:

```
configure terminal
interface serial 3/0/0
 ip route-cache flow
 exit
 ip flow-export 1.1.15.1 0 version 5 peer-as
 exit
 clear ip flow stats
```

PART IV

Tag Switching

CHAPTER 9

Tag Switching Overview

Tag Switching combines the performance and capabilities of Layer 2 (data link layer) switching with the proven scalability of Layer 3 (network layer) routing. It enables service providers to meet challenges brought about by explosive growth and provides the opportunity for differentiated services without necessitating the sacrifice of existing infrastructure. The Tag Switching architecture is remarkable for its flexibility. Data can be transferred over any combination of Layer 2 technologies, support is offered for all Layer 3 protocols, and scaling is possible well beyond anything offered in today's networks.

Specifically, Tag Switching can efficiently enable the delivery of IP services over an ATM switched network. It supports the creation of different routes between a source and a destination on a purely router-based Internet backbone. Service providers who use Tag Switching can save money and increase revenue and productivity.

Tag Switching offers the following benefits:

- IP over ATM scalability—Enables service providers to keep up with Internet growth

- IP services over ATM—Brings Layer 2 benefits to Layer 3, such as traffic engineering capability

- Standards—Supports multivendor solutions

- Architectural flexibility—Offers choice of ATM or router technology, or a mix of both

Tag Functions

In conventional Layer 3 forwarding, as a packet traverses the network, each router extracts all the information relevant to forwarding the packet from the Layer 3 header. This information is then used as an index for a routing table lookup to determine the packet's next hop.

In the most common case, the only relevant field in the header is the destination address field, but in some cases other header fields may also be relevant. As a result, the header analysis must be done independently at each router through which the packet passes, and a complicated lookup must also be done at each router.

In Tag Switching, the analysis of the Layer 3 header is done just once. The Layer 3 header is then mapped into a fixed-length, unstructured value called a *tag*.

Many different headers can map to the same tag, as long as those headers always result in the same choice of next hop. In effect, a tag represents a *forwarding equivalence class*—that is, a set of packets that, however different they may be, are indistinguishable to the forwarding function.

The initial choice of tag need not be based exclusively on the contents of the Layer 3 header; it can also be based on policy. This allows forwarding decisions at subsequent hops to be based on policy as well.

Once a tag is chosen, a short tag header is put at the front of the Layer 3 packet so that the tag value can be carried across the network with the packet. At each subsequent hop, the forwarding decision can be made simply by looking up the tag. There is no need to reanalyze the header. Since the tag is a fixed length and unstructured value, looking it up is fast and simple.

Distribution of Tag Bindings

Each *Tag Switching router* (TSR) makes an independent, local decision as to which tag value is used to represent which forwarding equivalence class. This association is known as a *tag binding*. Each TSR informs its neighbors of the tag bindings it has made. This is done by means of the Tag Distribution Protocol (TDP).

When a tagged packet is being sent from TSR A to a neighboring TSR B, the tag value carried by the packet is the tag value that TSR B assigned to represent the packet's forwarding equivalence class. Thus the tag value changes as the packet travels through the network.

Tag Switching and Routing

A tag represents a forwarding equivalence class, but it does not represent a particular path through the network. In general, the path through the network continues to be chosen by the existing Layer 3 routing algorithms such as OSPF, Enhanced IGRP, and BGP. That is, at each hop when a tag is looked up, the next hop chosen is determined by the dynamic routing algorithm.

Tag Switching and Traffic Engineering

In conventional Layer 3 routing, network topologies frequently include multiple paths between two points, but the normal routing procedure is to select a single path as the Layer 3 route between two points, regardless of the load on the links that implement the path. As a consequence, some links are congested and some are underused.

Traffic engineering provides a way to override routing protocols across multiple routers. It gives you the ability to direct selected traffic over specific paths in the network in order to efficiently use network resources and provide different levels of service.

To engineer your network traffic, you follow a two-step process. First, you define a sequence of links between two routers. Tag Switching is used to tunnel packets between the two routers over these links. The links collectively form a *Tag Switched path* (TSP) *tunnel*, which defines a traffic engineering path. Second, you select the traffic that you want forwarded on to the tunnel.

Traffic Engineering Tunnels and Filters

The traffic to be engineered is specified by a traffic engineering filter. The filter is associated with a TSP tunnel using a traffic engineering path.

The router at the head of the tunnel arranges that packets matching the filter be injected into the tunnel rather than being forwarded to their Layer 3 next hop. Injection consists simply of sending the packet to the first hop in the tunnel with a tag that causes that first hop to send the packet to the second hop of the tunnel, and so on.

For the initial release of traffic engineering, the only supported filtering is by "egress address." This filter matches traffic whose destination or BGP next hop is "address."

Multiple tunnels with different preferences can be specified for a single filter. A preference is an option you can select among multiple candidate routes for a filter, with the lower-valued preference being more desirable. The most preferred of the acceptable tunnels is used for the traffic.

A loop prevention algorithm operates to ensure that a tunnel is not used for traffic that might loop back to the head of the tunnel.

Traffic Engineering Tunnel Configuration

Configuration and the initiation of the tunnel are controlled by the *headend* (transmit end) router. Per-tunnel configuration of other routers is unnecessary.

Routers create and maintain the *traffic engineering tunnels* based on information you enter through the command-line interface (CLI). See Chapter 11, "Tag Switching Commands," for more information.

CHAPTER 10

Configuring Tag Switching

This chapter describes three sample cases where Tag Switching is configured on Cisco 7500/7200 series routers. These cases show the levels of control possible when Tag Switching is deployed in a network.

Table 10-1 lists the cases, including the steps to perform Tag Switching and their corresponding Cisco IOS CLI commands.

Table 10-1 *Tag Switching—Levels of Control*

This Case	Describes
Case 1—Enable Tag Switching incrementally in a network	The steps necessary for incrementally deploying Tag Switching through a network, assuming that packets to all destination prefixes should be tag switched.
Case 2—Route tagged packets to Network A only	The mechanism by which Tag Switching can be restricted, such that packets are tag switched to only a subset of destinations.
Case 3—Limit tag distribution on a Tag Switching network	The mechanisms for further controlling the distribution of a tag within a network.

For more information about the IOS CLI commands, see Chapter 11, "Tag Switching Commands."

Figure 10-1 shows a router-only Tag Switching network with Ethernet interfaces. The following sections outline the procedures for configuring Tag Switching and displaying Tag Switching information in a network based on the topology shown in Figure 10-1.

NOTE Ethernet interfaces are shown in Figure 10-1, but any of the interfaces that are supported could be used instead. ATM interfaces operating as TC-ATM interfaces are the exception to this statement.

Figure 10-1 *A Router-Only Tag Switching Network with Ethernet Interfaces*

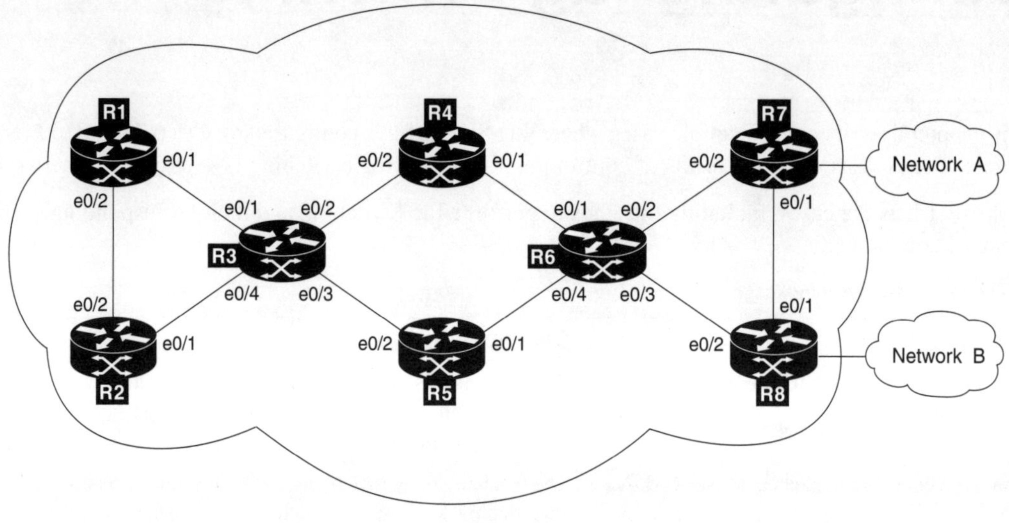

Case 1—Enable Tag Switching Incrementally in a Network

In the first case, assume that you want to deploy Tag Switching incrementally throughout a network of routers, but that you do not want to restrict which destination prefixes are tag switched. For a description of the commands listed in these cases, see Chapter 11, "Tag Switching Commands."

To enable Tag Switching incrementally in a network, perform these steps and enter the commands in router configuration mode (refer to Figure 10-1):

Step	Command	Purpose
1	**At R1:**	Enables Tag Switching between R1 and R3.
	Router# **configuration terminal** Router(config)# **ip cef distributed** Router(config)# **tag-switching advertise-tags** Router(config)# **interface e0/1** Router(config-if)# **tag-switching ip** Router(config-if)# **exit**	In order to configure distributed VIP Tag Switching, you must configure distributed CEF switching. Enter the **ip cef distributed** command on all routers.
	At R3:	
	Router# **configuration terminal** Router(config)# **ip cef distributed** Router(config)# **tag-switching advertise-tags** Router(config)# **interface e0/1** Router(config-if)# **tag-switching ip**	

Step	Command	Purpose
2	At R3: Router(config)# **interface e0/2** Router(config-if)# **tag-switching ip** Router(config-if)# **exit** At R4: Router# **configuration terminal** Router(config)# **ip cef distributed** Router(config)# **tag-switching advertise-tags** Router(config)# **interface e0/2** Router(config-if)# **tag-switching ip** Router(config-if)# **exit**	Enables Tag Switching between R3 and R4.

After you perform these steps, R1 applies tags to packets that are forwarded through interface e0/1, with a next hop to R3.

Tag switching can be enabled throughout the rest of the network by the repetition of steps 1 and 2 as appropriate on other routers until all routers and interfaces are enabled for Tag Switching. See the example in the section "Enabling Tag Switching Incrementally in a Network."

Case 2—Route Tagged Packets to Network A Only

In the second case, assume that you want to enable Tag Switching for a subset of destination prefixes. This option might be used to test Tag Switching across a large network. In this case, you would configure the system so that only a small number of destinations (for example, internal test networks) is tag switched without the majority of traffic being affected.

Perform the following steps at each router in the network in router configuration mode (refer to Figure 10-1).

Step	Command	Purpose
1	Router(config)# **access-list-1 permit A** (Enter the actual network address and netmask in place of **permit A**. For example, **access-list 1 permit 192.5.34. 0 0.0.0.255**.)	Limits tag distribution by using access lists.
2	Router(config)# **tag-switching advertise-tags for 1**	Instructs the router to advertise for Network A only to all adjacent tag switch routers. Any tags for other destination networks that the router may have distributed before this step are withdrawn.

Case 3—Limit Tag Distribution on a Tag Switching Network

The third case demonstrates the full control that is available to you in determining the destination prefixes and paths for which Tag Switching is enabled.

Configure the routers so that packets addressed to Network A are tagged, all other packets are untagged, and only links R1-R3, R3-R4, R4-R6, and R6-R7 carry tagged packets addressed to Network A. For example, suppose the normally routed path for packets arriving at R1 addressed to Network A or Network B is R1-R3-R5-R6-R7. A packet addressed to Network A would flow tagged on links R1-R3 and R6-R7, and untagged on links R3-R5 and R5-R6. A packet addressed to Network B would follow the same path, but would be untagged on all links.

Assume that at the outset the routers are configured so that packets addressed to Network A are tagged and all other packets are untagged (as at the completion of Case 2).

Use the **tag-switching advertise-tags** command and access lists to limit tag distribution. Specifically, you need to configure routers R2, R5, and R8 to distribute no tags to other routers. This ensures that no other routers will send tagged packets to any of those three. You also need to configure routers R1, R3, R4, R6, and R7 to distribute tags only for Network A and to distribute them only to the appropriate adjacent router; that is, R3 distributes its tag for Network A only to R1, R4 only to R3, and so on.

To limit tag distribution on a Tag Switching network, perform these steps in router configuration mode.

Step	Command	Purpose
1	Router(config)# **no tag-switching advertise-tags**	Configures R2 to distribute no tags.
2	Router(config)# **no tag-switching advertise-tags**	Configures R5 to distribute no tags.
3	Router(config)# **no tag-switching advertise-tags**	Configures R8 to distribute no tags.
4	Router(config)# **access-list 2 permit R1** Router(config)# **no tag-switching advertise-tags for 1** Router(config)# **tag-switching advertise-tags for 1 to 2** Router(config)# **exit** (Enter the actual network address and netmask in place of **permit R1**. For example, **access-list 1 permit 192.5.34.0 0.0.0.255**.)	Configures R3 by defining an access list and by instructing the router to distribute tags for the networks permitted by access list 1 (created as part of Case 2) to the routers permitted by access list 2. The **access list 2 permit R1** command permits R1 and denies all other routers.
5	Router(config)# **access-list 1 permit A** Router(config)# **access-list 2 permit R1** Router(config)# **tag-switching advertise-tags for 1 to 2** Router(config)# **exit** (Enter the actual network address and netmask in place of **permit A** and **permit R1**. For example, **access-list 1 permit 192.5.34.0 0.0.0.255**.)	Configures R3.

Step	Command	Purpose
6	Router(config)# **access-list 1 permit A** Router(config)# **access-list 2 permit R3** Router(config)# **tag-switching advertise-tags for 1 to 2** Router(config)# **exit** (Enter the actual network address and netmask in place of **permit A** and **permit R3**. For example, **access-list 1 permit 192.5.34.0 0.0.0.255**.)	Configures R4.
7	Router(config)# **access-list 1 permit A** Router(config)# **access-list 2 permit R4** Router(config)# **tag-switching advertise-tags for 1 to 2** Router(config)# **exit** (Enter the actual network address and netmask in place of **permit A** and **permit R4**. For example, **access-list 1 permit 192.5.34.0 0.0.0.255**.)	Configures R6.
8	Router(config)# **access-list 1 permit A** Router(config)# **access-list 2 permit R6** Router(config)# **tag-switching advertise-tags for 1 to 2** Router(config)# **exit** (Enter the actual network address and netmask in place of **permit A** and **permit R6**. For example, **access-list 1 permit 192.5.34.0 0.0.0.255**.)	Configures R7.

Traffic Engineering

This section describes two sample cases supported by traffic engineering. These cases show how you can engineer traffic across a path in the network and establish a backup route for that traffic engineered path (see Table 10-2).

In both cases, the assumption is made that traffic from R1 and R2 (in Figure 10-2), which is intended for R11, would be directed by Layer 3 routing along the "upper" path R3-R4-R7-R10-R11.

Table 10-2 *Sample Traffic Engineering Cases*

This case	Describes
Case 1—Engineer traffic across a path	The steps necessary to engineer traffic across the "middle" path R3-R5-R8 (see Figure 10-2).
Case 2—Establish a backup path	The steps necessary for establishing a backup traffic engineering route for the engineered traffic for Case 1.

Figure 10-2 shows a router-only Tag Switching network with traffic engineered paths.

Figure 10-2 *Sample Tag Switching Network with Traffic Engineered Paths*

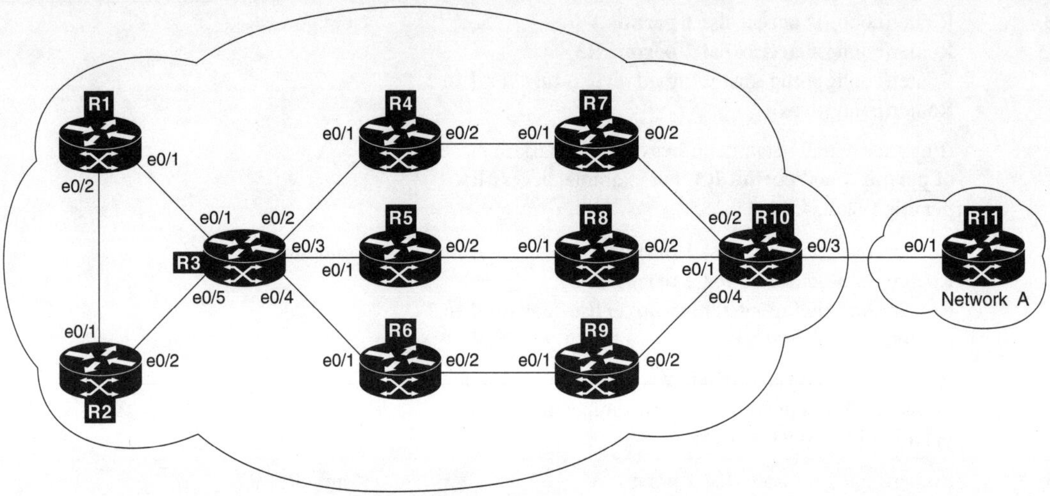

Case 1—Engineer Traffic Across a Path.

The following table lists the configuration commands you need to engineer traffic across the "middle" path R3-R5-R8 by building a tunnel R1-R3-R5-R8-R10, without affecting the path taken by traffic from R2 (see Figure 10-2).

To engineer traffic across a path, perform the following steps in router configuration mode:

Step	Command	Purpose
1	At R1: Router(config)# **ip cef distributed** Router(config)# **tag-switching tsp-tunnels** Router(config)# **interface e0/1** Router(config-if)# **tag-switching tsp-tunnels** Router(config-if)# **exit** At R3: Router(config)# **ip cef distributed** Router(config)# **tag-switching tsp-tunnels** Router(config)# **interface e0/1** Router(config-if)# **tag-switching tsp-tunnels** Router(config-if)# **exit** Router(config)# **interface e0/3** Router(config-if)# **tag-switching tsp-tunnels** Router(config-if)# **exit**	Configures support for TSP tunnel signaling along the path. In order to configure distributed VIP Tag Switching, you must configure distributed CEF switching. Enter the **ip cef distributed** command on all routers. Note: To configure a Cisco 7200 series router, enter **ip cef**. To configure a Cisco 7500 series router, enter **ip cef distributed.**

Step	Command	Purpose
1	At R5 and R8: Router(config)# **ip cef distributed** Router(config)# **tag-switching tsp-tunnels** Router(config)# **interface e0/1** Router(config-if)# **tag-switching tsp-tunnels** Router(config-if)# **exit** Router(config)# **interface e0/2** Router(config-if)# **tag-switching tsp-tunnels** Router(config-if)# **exit** At R10: Router(config)# **ip cef distributed** Router(config)# **tag-switching tsp-tunnels** Router(config)# **interface e0/1** Router(config-if)# **tag-switching tsp-tunnels** Router(config-if)# **exit**	
2	At R1: Router(config)# **interface tunnel 2003** Router(config-if)# **ip unnumbered e0/1** Router(config-if)# **tunnel mode tag-switching** Router(config-if)# **tunnel tsp-hop 1 10.10.0.103** Router(config-if)# **tunnel tsp-hop 2 10.11.0.105** Router(config-if)# **tunnel tsp-hop 3 10.12.0.108** Router(config-if)# **tunnel tsp-hop 4 10.13.0.110 lasthop** Router(config-if)# **exit**	Configures a TSP tunnel at the headend. (IP address of R3:e0/1) (IP address of R5:e0/1) (IP address of R8:e0/1) (IP address of R10:e0/1)
3	At R1: Router(config)# **router traffic-engineering** Router(config)# **traffic-engineering filter 1 egress 10.14.0.111 255.255.255.255**	Configures the traffic engineering filter to classify the traffic to be routed. The filter selects all traffic where the autonomous system (AS) egress router is 10.14.0.111 (10.14.0.111 is the IP address of R11:e0/1).
4	At R1: Router(config)# **router traffic-engineering** Router(config)# **traffic-engineering route 1 tunnel 2003**	Configures the traffic engineering route to send the engineered traffic down the tunnel.

Case 2—Establish a Backup Path

Case 2 involves establishing a backup traffic engineering route for the engineered traffic for Case 1. This backup route uses the "lower" path. The backup route uses a tunnel R1-R3-R6 and relies on Layer 3 routing to deliver the packet from R6 to R11.

To set up a traffic engineering backup path (assuming that Case 1 steps have been performed), follow these steps in router configuration mode:

Step	Command	Purpose
1	At R6: Router(config)# **ip cef distributed** Router(config)# **tag-switching tsp-tunnels** Router(config)# **interface e0/1** Router(config-if)# **tag-switching tsp-tunnels** Router(config-if)# **exit** At R3: Router(config)# **ip cef distributed** Router(config)# **tag-switching tsp-tunnels** Router(config)# **interface e0/4** Router(config-if)# **tag-switching tsp-tunnels** Router(config-if)# **exit**	Enables TSP tunnel signalling along the path (where such signalling is not already enabled).
2	At R1: Router(config)# **interface tunnel 2004** Router(config-if)# **ip unnumbered e0/1** Router(config-if)# **tunnel mode tag-switching** Router(config-if)# **tunnel tsp-hop 1 10.10.0.103** Router(config-if)# **tunnel tsp-hop 2 10.21.0.106 lasthop** Router(config-if)# **exit**	Configures the TSP tunnel at the headend. (IP address of R3:e0/1) (IP address of R6:e0/1)
3	At R1: Router(config)# **router traffic-engineering** Router(config)# **traffic-engineering route 1 tunnel 2004 pref 200**	Configures the traffic engineering route to send the engineered traffic down the tunnel if the middle path (Case 1 route) is unavailable.

Configuration Examples

This section provides sample configurations for the Cisco 7500/7200 series routers. It contains the following sections:

- Enabling Tag Switching Incrementally in a Network

- Enabling Tag Switching for a Subset of Destination Prefixes

- Selecting the Destination Prefixes and Paths

- Displaying Tag Switching TDP Binding Information

- Displaying Tag Switching Forwarding Table Information

- Displaying Tag Switching Interface Information

- Displaying Tag Switching TDP Neighbor Information

- Enabling TSP Tunnel Signalling

- Configuring a TSP Tunnel

- Displaying the TSP Tunnel Information

- Configuring a Traffic Engineering Filter and Route

- Displaying Traffic Engineering Configuration Information

Enabling Tag Switching Incrementally in a Network

The following example shows you how to configure Tag Switching incrementally throughout a network of routers. You enable Tag Switching first between one pair of routers (in this case, R1 and R3, as shown in Figure 10-1) and add routers step-by-step until every router in the network is Tag Switching enabled.

```
router-1# configuration terminal
router-1(config)# ip cef distributed
router-1(config)# tag-switching ip
router-1(config)# interface e0/1
router-1(config-if)# tag-switching ip
router-1(config-if)# exit
router-1(config)#
router-3# configuration terminal
router-3(config)# ip cef distributed
router-3(config)# tag-switching ip
router-3(config)# interface e0/1
router-3(config-if)# tag-switching ip
router-3(config-if)# exit
router-3(config)#
```

Enabling Tag Switching for a Subset of Destination Prefixes

The following example shows the commands you enter at each of the routers to enable Tag Switching for only a subset of destination prefixes (refer to Figure 10-1).

```
Router(config)# access-list-1 permit A
Router(config)# tag-switching advertise-tags for 1
```

Selecting the Destination Prefixes and Paths

The following example shows the commands you enter to configure the routers to select the destination prefixes and paths for which Tag Switching is enabled. When you configure R2, R5, and R8 to distribute no tags to other routers, you ensure that no routers send them tagged packets. You also need to configure routers R1, R3, R4, R6, and R7 to distribute tags only for Network A and only to the applicable adjacent router. This configuration ensures that R3 distributes its tag for Network A only to R1, R4 only to R3, R6 only to R4, and R7 only to R6 (refer to Figure 10-1).

```
router-2(config)# no tag-switching advertise-tags
router-5(config)# no tag-switching advertise-tags
router-8(config)# no tag-switching advertise-tags
router-1(config)# access-list permit R1
router-1(config)# no tag-switching advertise-tags for 1
router-1(config)# tag-switching advertise-tags for 1 to 2
router-1(config)# exit

router-3# access-list 1 permit A
router-3# access-list 2 permit R1
router-3# tag-switching advertise-tags for 1 to 2
router-3# exit

router-4# access-list 1 permit A
router-4# access-list 2 permit R3
router-4# tag-switching advertise-tags for 1 to 2
router-4# exit
router-6# access-list 1 permit A
router-6# access-list 2 permit R4
router-6# tag-switching advertise-tags for 1 to 2
router-6# exit
router-7# access-list 1 permit A
router-7# access-list 2 permit R6
router-7# tag-switching advertise-tags for 1 to 2
router-7# exit
```

Displaying Tag Switching TDP Binding Information

Use the **show tag-switching tdp bindings** command to display the contents of the Tag Information
Base (TIB). The display can show the entire database or can be limited to a subset of entries, based on
prefix, input or output tag values or ranges, and/or the neighbor advertising the tag.

NOTE   This command displays downstream mode bindings. For tag VC bindings, see the **show
tag-switching atm-tdp bindings** command.

```
Router# show tag-switching tdp bindings

Matching entries:
  tib entry: 10.92.0.0/16, rev 28
       local binding:  tag: imp-null(1)
       remote binding: tsr: 172.27.32.29:0, tag: imp-null(1)
  tib entry: 10.102.0.0/16, rev 29
       local binding:  tag: 26
       remote binding: tsr: 172.27.32.29:0, tag: 26
  tib entry: 10.105.0.0/16, rev 30
       local binding:  tag: imp-null(1)
       remote binding: tsr: 172.27.32.29:0, tag: imp-null(1)
```

```
tib entry: 10.205.0.0/16, rev 31
      local binding:  tag: imp-null(1)
      remote binding: tsr: 172.27.32.29:0, tag: imp-null(1)
tib entry: 10.211.0.7/32, rev 32
      local binding:  tag: 27
      remote binding: tsr: 172.27.32.29:0, tag: 28
tib entry: 10.220.0.7/32, rev 33
      local binding:  tag: 28
      remote binding: tsr: 172.27.32.29:0, tag: 29
tib entry: 99.101.0.0/16, rev 35
      local binding:  tag: imp-null(1)
      remote binding: tsr: 172.27.32.29:0, tag: imp-null(1)
tib entry: 100.101.0.0/16, rev 36
      local binding:  tag: 29
      remote binding: tsr: 172.27.32.29:0, tag: imp-null(1)
tib entry: 171.69.204.0/24, rev 37
      local binding:  tag: imp-null(1)
      remote binding: tsr: 172.27.32.29:0, tag: imp-null(1)
tib entry: 172.27.32.0/22, rev 38
      local binding:  tag: imp-null(1)
      remote binding: tsr: 172.27.32.29:0, tag: imp-null(1)
tib entry: 210.10.0.0/16, rev 39
      local binding:  tag: imp-null(1)
tib entry: 210.10.0.8/32, rev 40
      remote binding: tsr: 172.27.32.29:0, tag: 27
```

Displaying Tag Switching Forwarding Table Information

Use the **show tag-switching forwarding-table** command to display the contents of the Tag Forwarding Information Base (TFIB). The TFIB lists the tags, output interface information, prefix or tunnel associated with the entry, and number of bytes received with each incoming tag. A request can show the entire TFIB or can be limited to a subset of entries. A request can also be restricted to selected entries in any of the following ways:

● A single entry associated with a given incoming tag

● Entries associated with a given output interface

● Entries associated with a given next hop

● A single entry associated with a given destination

● A single entry associated with a given tunnel having the current node as an intermediate hop

```
Router# show tag-switching forwarding-table

Local  Outgoing     Prefix         Bytes tag  Outgoing     Next Hop
tag    tag or VC    or Tunnel Id   switched   interface
26     Untagged     10.253.0.0/16  0          Et4/0/0      172.27.32.4
28     1/33         10.15.0.0/16   0          AT0/0.1      point2point
29     Pop tag      10.91.0.0/16   0          Hs5/0        point2point
       1/36         10.91.0.0/16   0          AT0/0.1      point2point
```

```
30   32              10.250.0.97/32   0    Et4/0/2    10.92.0.7
     32              10.250.0.97/32   0    Hs5/0      point2point
34   26              10.77.0.0/24     0    Et4/0/2    10.92.0.7
     26              10.77.0.0/24     0    Hs5/0      point2point
35   Untagged   [T] 10.100.100.101/32 0   Tu301      point2point
36   Pop tag         168.1.0.0/16     0    Hs5/0      point2point
     1/37            168.1.0.0/16     0    AT0/0.1    point2point

[T]     Forwarding through a TSP tunnel.
        View additional tagging info with the 'detail' option
```

Displaying Tag Switching Interface Information

Use the **show tag-switching interfaces** command to show information about the requested interface or about all interfaces on which Tag Switching is enabled. The per-interface information includes the interface name and indications as to whether IP Tag Switching is enabled and operational.

```
Router# show tag-switching interfaces

Interface          IP    Tunnel  Operational
Hssi3/0            Yes   Yes     No
ATM4/0.1           Yes   Yes     Yes       (ATM tagging)
Ethernet5/0/0      No    Yes     Yes
Ethernet5/0/1      Yes   No      Yes
Ethernet5/0/2      Yes   No      No
Ethernet5/0/3      Yes   No      Yes
Ethernet5/1/1      Yes   No      No
```

The following shows sample output from the **show tag-switching interfaces** command when you specify **detail:**

```
Router# show tag-switching interface detail

Interface Hssi3/0:
        IP tagging enabled
        TSP Tunnel tagging enabled
        Tagging not operational
        MTU = 4470
Interface ATM4/0.1:
        IP tagging enabled
        TSP Tunnel tagging enabled
        Tagging operational
        MTU = 4470
        ATM tagging: Tag VPI = 1, Control VC = 0/32
Interface Ethernet5/0/0:
        IP tagging not enabled
        TSP Tunnel tagging enabled
        Tagging operational
        MTU = 1500
Interface Ethernet5/0/1:
        IP tagging enabled
        TSP Tunnel tagging not enabled
        Tagging operational
        MTU = 1500
```

```
Interface Ethernet5/0/2:
        IP tagging enabled
        TSP Tunnel tagging not enabled
        Tagging not operational
        MTU = 1500
Interface Ethernet5/0/3:
        IP tagging enabled
        TSP Tunnel tagging not enabled
        Tagging operational
        MTU = 1500
```

Displaying Tag Switching TDP Neighbor Information

Use the **show tag-switching tdp neighbors** command to display the status of Tag Distribution Protocol (TDP) sessions. The neighbor information branch can have information about all TDP neighbors or can be limited to the neighbor with a specific IP address or TDP identifier, or to TDP neighbors known to be accessible over a specific interface.

```
Router# show tag-switching tdp neighbors

Peer TDP Ident: 10.220.0.7:1; Local TDP Ident 172.27.32.29:1
        TCP connection: 10.220.0.7.711 - 172.27.32.29.11029
        State: Oper; PIEs sent/rcvd: 17477/17487; Downstream on demand
Up time: 01:03:00
TDP discovery sources:
        ATM0/0.1
Peer TDP Ident: 210.10.0.8:0; Local TDP Ident 172.27.32.29:0
        TCP connection: 210.10.0.8.11004 - 172.27.32.29.711
        State: Oper; PIEs sent/rcvd: 14656/14675; Downstream;
Up time: 2d5h
        TDP discovery sources:
          Ethernet4/0/1
          Ethernet4/0/2
          POS6/0/0
        Addresses bound to peer TDP Ident:
          99.101.0.8      172.27.32.28     10.105.0.8     10.92.0.8
          10.205.0.8      210.10.0.8
```

Enabling TSP Tunnel Signalling

The following example shows you how to configure support for Tag Switched path (TSP) tunnel signalling along a path and on each interface crossed by one or more tunnels:

```
Router(config)# ip cef distributed
Router(config)# tag-switching tsp-tunnels
Router(config)# interface e0/1
Router(config-if)# tag-switching tsp-tunnels
Router(config-if)# interface e0/2
Router(config-if)# tag-switching tsp-tunnels
Router(config-if)# exit
```

Configuring a TSP Tunnel

The following example shows you how to set the encapsulation of the tunnel to Tag Switching and how to define hops in the path for the TSP.

Follow these steps to configure a two-hop tunnel, hop 0 being the headend router. For hops 1 and 2, you specify the IP addresses of the incoming interfaces for the tunnel. The tunnel interface number is arbitrary, but must be less than 65,535:

```
Router(config)# interface tunnel 2003
Router(config-if)# tunnel mode tag-switching
Router(config-if)# tunnel tsp-hop 1 10.10.0.12
Router(config-if)# tunnel tsp-hop 2 10.50.0.24 lasthop
Router(config-if)# exit
```

To shorten the previous path, you delete a hop by entering the following commands:

```
Router(config)# interface tunnel 2003
Router(config-if)# no tunnel tsp-hop 2
Router(config-if)# tunnel tsp-hop 1 10.10.0.12 lasthop
Router(config-if)# exit
```

Displaying the TSP Tunnel Information

Use the **show tag-switching tsp tunnels** command to display information about the configuration and status of selected tunnels.

```
Router# show tag-switching tsp-tunnels

Signalling Summary:
          TSP Tunnels Process:        running
          RSVP Process:               running
          Forwarding:                 enabled

TUNNEL ID   DESTINATION      STATUS     CONNECTION
10.106.0.6.200310.2.0.12up  up
```

Configuring a Traffic Engineering Filter and Route

The following example shows you how to configure the traffic engineering routing process, a traffic engineering filter, and a traffic engineering route for that filter over a TSP-encapsulated tunnel.

```
Router(config)# router traffic-engineering
Router(config-router)# traffic-engineering filter 5 egress 83.0.0.1 255.255.255.255
Router(config-router)# traffic-engineering route 5 tunnel 5
```

Displaying Traffic Engineering Configuration Information

Use the **show ip traffic-engineering configuration** command to display information about the configured traffic engineering filters and routes. The following is sample output from the **show ip traffic-engineering configuration detail** command.

```
Router# show ip traffic-engineering configuration detail

Traffic Engineering Configuration
    Filter 5: egress 44.0.0.0/8, local metric: ospf-0/1
        Tunnel5 route installed
          interface up, route enabled, preference 1
          loop check on, passing, remote metric: connected/0
    Filter 6: egress 43.0.0.1/32, local metric: ospf-300/3
        Tunnel7 route installed
          interface up, route enabled, preference 50
          loop check on, passing, remote metric: ospf-300/2
        Tunnel6 route not installed
          interface up, route enabled, preference 75
          loop check on, passing, remote metric: connected/0
```

CHAPTER 11

Tag Switching Commands

This chapter documents commands used to configure Tag Switching features in Cisco IOS software. For guidelines on configuring Tag Switching features, refer to Chapter 10, "Configuring Tag Switching."

NOTE   Beginning with Cisco IOS Release 11.3, all commands supported on the Cisco 7500 series routers are also supported on Cisco 7000 series routers.

---

router traffic-engineering

To configure the traffic engineering routing process, use the **router traffic-engineering** global configuration command. To turn off the traffic engineering routing process and delete any associated configuration, use the **no** form of this command.

> **router traffic-engineering**
> **no router traffic-engineering**

Syntax Description

This command has no arguments or keywords.

Default

Traffic engineering process is disabled.

Command Mode

Global configuration

Usage Guidelines

This command first appeared in Cisco IOS Release 11.1 CT.

Example

In the following example, configuration is provided for a traffic engineering routing process, a traffic engineering filter, and a traffic engineering route for that filter over a TSP tunnel:

```
router traffic-engineering
  traffic-engineering filter 5 egress 83.0.0.1 255.255.255.255
  traffic-engineering route 5 tunnel 5
```

Related Commands

You can search online at www.cisco.com to find documentation of related commands.

show ip traffic-engineering
traffic-engineering filter
traffic-engineering route

show ip traffic-engineering

To display information about the traffic engineering configuration and metric information associated with it, use the **show ip traffic-engineering** privileged EXEC command.

> **show ip traffic-engineering [metrics [detail]]**

Syntax	Description
metrics	(Optional) Displays metric information associated with traffic engineering.
detail	(Optional) Displays information in long form.

Command Mode

Privileged EXEC

Usage Guidelines

This command first appeared in Cisco IOS Release 11.1 CT.

The goal of the loop prevention algorithm is that traffic should not be sent down the tunnel if there is a possibility that, after leaving the tunnel, steady state routing will route the traffic back to the head of the tunnel.

The approach of the loop prevention algorithm is to compare the Layer 3 routing distance to the egress from the tunnel tailend and tunnel headend. The loop check passes only if the tunnel tail is closer to the egress than the tunnel head is.

The loop prevention algorithm allows you to use the tunnel for a route if one the following cases applies:

● Given that the two ends of the tunnel are routing to the egress using the same dynamic protocol in the same area, the Layer 3 routing distance from the tailend to the egress is less than the Layer 3 routing distance from the headend to the egress.

● The route to the egress is directly connected at the tunnel tailend router, but not at the tunnel headend router.

● The egress is unreachable from the tunnel headend router, but is reachable from the tunnel tailend router.

The loop prevention algorithm prevents you from using the tunnel for a given egress in all other cases. In particular, this occurs when

● The routers at the ends of the tunnel get their route to the egress from different dynamic routing protocols.

● The routing protocols at the two ends of the tunnel route to the egress through different areas.

● The two ends each use a static route to the egress.

● The tunnel headend router's route to the egress is a connected route.

● The egress is unreachable from the tunnel tailend router.

Devices request metrics via a TDP adjacency. The display output shows detailed metric information.

The metric information includes a metric type (shown as routing_protocol/routing_protocol_subtype) and a metric value.

The routing protocol is one of the following:

    OSPF
    IS-IS
    EIGRP
    Connected
    Static
    Other (some other routing protocol)

The routing protocol subtype is specific to each routing protocol.

Sample Display

The following is sample output from the **show ip traffic-engineering metrics detail** command:

```
Router# show ip traffic-engineering metrics detail

Metrics requested BY this device
 Prefix 43.0.0.1/32
  TDP id 2.2.2.2:0, metric: connected/0
    type request, flags metric-received, rev 6, refcnt 1
  TDP id 4.4.4.4:0, metric: ospf-300/2
    type request, flags metric-received, rev 7, refcnt 1
 Prefix 44.0.0.0/8
  TDP id 18.18.18.18:0, metric: connected/0
    type request, flags metric-received, rev 1, refcnt 1
Metrics requested FROM this device
 Prefix 36.0.0.0/8
  TDP id 18.18.18.18:0, metric: connected/0
    type advertise, flags none, rev 1, refcnt 1
```

Table 11-1 defines the fields displayed in the first three lines of the output from the **show ip traffic-engineering metrics detail** command.

Table 11-1 *show ip traffic-engineering metrics detail* Field Descriptions

Field	Description
Prefix	Destination network and mask.
TDP id	The TDP identifier of the TDP peer device at the other end of the tunnel. The TDP peer device advertises these metrics to this neighbor.
metric	The routing protocol and metric within that protocol for the prefix in question.
type	For metrics being requested by this device, the type is either request or release. For metrics being requested from this device, the type is advertise.
flags	For metrics being requested by this device metric-received indicates that the other end has responded with a metric value. For metrics being requested from this device, response-pending indicates that the metric value has not yet been sent to the requester.
rev	An internal identifier for the metric request or advertisement. The rev number is assigned when the request/advertisement is created. The rev number is updated if the local information for the metric changes.
refcnt	For a metric of type request, the number of traffic engineering routes interested in this metric value. Otherwise, refcnt is 1.

Related Commands

You can search online at www.cisco.com to find documentation of related commands.

router traffic-engineering
traffic-engineering filter
traffic-engineering route

show ip traffic-engineering configuration

To display information about configured traffic engineering filters and routes, use the **show ip traffic-engineering configuration** privileged EXEC command.

       show ip traffic-engineering configuration [*interface*] [*filter-number*] [**detail**]

Syntax	Description
interface	(Optional) Specifies an interface for which to display traffic engineering information.
filter-number	(Optional) A decimal value representing the number of the filter to display.
detail	(Optional) Displays command output in long form.

Command Mode

Privileged EXEC

Usage Guidelines

This command first appeared in Cisco IOS Release 11.1 CT.

The sample output can show all filters or can be limited by interface, filter number, or both.

Sample Display

The following is sample output from the **show ip traffic-engineering configuration detail** command:

```
Router# show ip traffic-engineering configuration detail

Traffic Engineering Configuration
    Filter 5: egress 44.0.0.0/8, local metric: ospf-0/1
        Tunnel5 route installed
          interface up, preference 1
          loop check on, passing, remote metric: connected/0
```

```
Filter 6: egress 43.0.0.1/32, local metric: ospf-300/3
    Tunnel7 route installed
       interface up, preference 50
       loop check on, passing, remote metric: ospf-300/2
    Tunnel6 route not installed
       interface up, preference 75
       loop check on, passing, remote metric: connected/0
```

Table 11-2 describes the fields displayed in the first four lines of the output from the **show ip traffic-engineering configuration detail** command.

Table 11-2 *show ip traffic engineering configuration detail Field Descriptions*

Field	Description
Filter	The configured filter identifier for the traffic engineering route.
egress	The prefix/mask configured with the filter local metric.
local metric	The local TSR's routing protocol and metric value for the egress prefix/mask.
Tunnel5	The tunnel for the traffic engineering route.
route installed/not installed	Indicates whether the route is installed in the forwarding tables (typically CEF and tag interface up/down).
interface	Indicates whether the tunnel interface for the traffic engineering route is up or down. The traffic engineering route is not installed if the tunnel interface is down.
preference	The configured administrative preference for the traffic engineering route.
loop check	Indicates whether the loop check has been configured on or off.
passing/failing	If the loop check is configured on, indicates whether the check is passing. The traffic engineering route is not installed if the loop check is configured on and is failing.
remote metric	The routing protocol and the metric within that protocol for the prefix in question, as seen by the TSR advertising the metric. As part of the loop check, a comparison is made between the remote metric and the local metric.

Related Commands

You can search online at www.cisco.com to find documentation of related commands.

show ip traffic-engineering routes

show ip traffic-engineering routes

To display information about the requested filters configured for traffic engineering, use the **show ip traffic-engineering routes** privileged EXEC command.

show ip traffic-engineering routes [*filter-number*] [**detail**]

Syntax / Description

Syntax	Description
filter-number	(Optional) A decimal value representing the number of the filter to display.
detail	(Optional) Display of command output in long form.

Command Mode

Privileged EXEC

Usage Guidelines

This command first appeared in Cisco IOS Release 11.1 CT. Requests can be limited to a specific filter.

Sample Display

The following is sample output from the **show ip traffic-engineering routes** command:

```
Router# show ip traffic-engineering routes

Installed traffic engineering routes:
Codes: T - traffic engineered route
T    43.0.0.1/32 (not override of routing table entry)
             is directly connected, 00:06:35, Tunnel7
T    44.0.0.0/8 (override of routing table entry)
             is directly connected, 01:12:39, Tunnel5
```

Table 11-3 describes the significant fields in the output from the **show ip traffic-engineering routes** command.

Table 11-3 *show ip traffic-engineering routes Field Descriptions*

Field	Description
T	Traffic engineering route.
43.0.0.1/32 (not override of routing table entry) is directly connected	Prefix/mask being routed. The routing table does not contain an entry for this prefix/mask.

Continues

Table 11-3 *show ip traffic-engineering routes Field Descriptions (Continued)*

Field	Description
00:06:35	The time since the route was installed (hours:minutes:seconds).
Tunnel7	The TSP tunnel for the route.

Related Commands

You can search online at www.cisco.com to find documentation of related commands.

show ip traffic-engineering configuration

show tag-switching atm-tdp bindings

To display the requested entries from the ATM TDP tag binding database, use the **show tag-switching atm-tdp bindings** privileged EXEC command. The ATM TDP database contains TIB entries for tag VCs on TC-ATM interfaces.

> **show tag-switching atm-tdp bindings** [*network* {*mask* | *length*}]
> [**local-tag** *vpi vci*][**remote-tag** *vpi vci*] [**neighbor** *interface*]

Syntax	Description
network	(Optional) Destination network number.
mask	Network mask in the form A.B.C.D (destination prefix).
length	Mask length (1 to 32).
local-tag *vpi vci*	(Optional) Select tag VC value assigned by this router.
remote-tag *vpi vci*	(Optional) Select tag values assigned by the other router.
neighbor *interface*	(Optional) Select tag values assigned by neighbor on a specified interface.

Command Mode

Privileged EXEC

Usage Guidelines

This command first appeared in Cisco IOS Release 11.1 CT.

The display output can show entries from the entire database, or it can be limited to a subset of entries based on prefix, VC tag value, and/or an assigning interface.

Sample Displays

The following is router sample output from the **show tag-switching atm-tdp bindings** command:

```
Router# show tag-switching atm-tdp bindings

Destination: 10.16.0.16/32
    Tailend Router ATM1/0.1 1/35 1/34 Active, VCD=2
Destination: 10.24.0.0/24
    Tailend Router ATM1/0.1 1/39 Active, VCD=3
Destination: 10.15.0.15/32
    Tailend Router ATM1/01 1/33 Active, VCD=4
Destination: 10.23.0 0/24
    Tailend Router ATM1/01 1/37 Active, VCD=5
```

Table 11-4 describes the significant fields in the output from the **show tag-switching atm-tdp bindings** command.

Table 11-4 *show tag-switching atm-tdp bindings Field Descriptions*

Field	Description
Destination:	Destination (network/mask)
Tailend Router	Types of VC. Options are
	Tailend—VC that terminates at this router
	Headend—VC that originates at this router
	Transit—VC that passes through a switch
ATM1/0.1	Interface.
1/35	VPI/VCI.
Active	TVC state:
	Active—Set up and working.
	Bindwait—Waiting for response.
	Remote Resource Wait—Waiting for resources (VPI/VCI space) to be available on the downstream device.
	Parent Wait—Transit VC input side waiting for output side to become active.
VCD=2	Virtual circuit descriptor number.

The following is ATM switch sample output from the **show tag-switching atm-tdp bindings** command:

```
Switch# show tag-switching atm-tdp bindings

Destination: 6.6.6.6/32
    Tailend Switch ATM0/0/3 1/34 Active -> Terminating Active
 Destination: 150.0.0.0/16
    Tailend Switch ATM0/0/3 1/35 Active -> Terminating Active
 Destination: 4.4.4.4/32
    Transit ATM0/0/3 1/33 Active -> ATM0/1/1 1/33 Active
```

Related Commands

You can search online at www.cisco.com to find documentation of related commands.

show tag-switching atm-tdp summary

show tag-switching atm-tdp capability

To display the ATM TDP tag capabilities, use the **show tag-switching atm-tdp capability** privileged EXEC command.

> **show tag-switching atm-tdp capability**

Syntax Description

This command has no arguments or keywords.

Command Mode

Privileged EXEC

Usage Guidelines

This command first appeared in Cisco IOS Release 11.1 CT.

Sample Display

The following example shows the display from the **show tag-switching atm-tdp capability** command:

```
Router# show tag-switching atm-tdp capability

VPI         VCI         Alloc    Odd/Even  VC Merge
ATM0/1/0    Range       Range              Scheme  Scheme   IN   OUT
  Negotiated  [100 - 101]  [33 - 1023]   UNIDIR              -    -
```

```
Local      [100 - 101]   [33 - 16383]   UNIDIR              EN    EN
Peer       [100 - 101]   [33 - 1023]    UNIDIR               -     -

           VPI           VCI            Alloc   Odd/Even  VC Merge
ATM0/1/1   Range         Range          Scheme  Scheme    IN    OUT
  Negotiated [201 - 202] [33 - 1023]    BIDIR              -     -
  Local    [201 - 202]   [33 - 16383]   UNIDIR  ODD        NO    NO
  Peer     [201 - 202]   [33 - 1023]    BIDIR   EVEN       -     -
```

Table 11-5 lists the significant fields in the display from the **show tag-switching atm-tdp capability** command.

Table 11-5 *show tag-switching atm-tdp capability Field Descriptions*

Field	Description
VPI Range	Minimum and maximum number of VPIs supported on this interface.
VCI Range	Minimum and maximum number of VCIs supported on this interface.
Alloc Scheme	UNIDIR—Unidirectional capability indicates that the peer device can, within a single VPI, support binding of the same VCI to different prefixes on different directions of the link.
	BIDIR—Bidirectional capability indicates that within a single VPI, a single VCI can appear in one binding only. In this case, one peer device allocates bindings in the even VCI space, and the other in the odd VCI space. The system with the lower TDP identifier will assign even-numbered VCIs.
	The negotiated allocation scheme is UNIDIR if and only if both peer devices have UNIDIR capability. Otherwise it is BIDIR.
Odd/Even Scheme	Indicates whether the local device or the peer device is assigning an odd- or even-numbered VCI when the negotiated scheme is BIDIR. It does not display any information when the negotiated scheme is UNIDIR.

Continues

Table 11-5 *show tag-switching atm-tdp capability Field Descriptions (Continued)*

Field	Description
VC Merge	Indicates the type of VC merge support on this interface.
	IN—Indicates input interface merge capability. IN accepts the following values:
	EN—The hardware interface supports VC merge and VC merge is enabled on the device.
	DIS—The hardware interface supports VC merge and VC merge is disabled on the device.
	NO—The hardware interface does not support VC merge.
	OUT—Indicates output interface merge capability. OUT accepts the same values as the input merge side.
	The VC merge capability is meaningful only on ATM switches. It is not negotiated.
Negotiated	Set of options that both TDP peer devices have agreed to share on this interface. For example, the VPI or VCI allocation on either peer device remains within the negotiated ranges.
Local	Options supported locally on this interface.
Peer	Options supported by the remote TDP peer device on this interface.

Related Commands

You can search online at www.cisco.com to find documentation of related commands.

tag-switching atm control-vc
tag-switching atm vc-merge
tag-switching atm vpi

show tag-switching atm-tdp summary

To display summary information on ATM tag bindings, use the **show tag-switching atm-tdp summary** privileged EXEC command.

show tag-switching atm-tdp summary

Syntax Description

This command has no arguments or keywords.

Command Mode

Privileged EXEC

Usage Guidelines

This command first appeared in Cisco IOS Release 11.1 CT.

Sample Display

The following is sample output from the **show tag-switching atm-tdp summary** command:

```
Router# show tag-switching atm-tdp summary

Total number of destinations: 788

TC-ATM bindings summary
interface   total    active    bindwait  local    remote    other
ATM0/0/0    594      592       1         296      298       1
ATM0/0/1    590      589       0         294      296       1
ATM0/0/2    1179     1178      0         591      588       1
ATM0/0/3    1177     1176      0         592      585       1
ATM0/1/0    1182     1178      4         590      588       0
Waiting for bind on ATM0/0/0 10.21.0.0/24
```

Table 11-6 describes the significant fields in the output from the **show tag-switching atm-tdp summary** command.

Table 11-6 *show tag-switching atm-tdp summary Field Descriptions*

Field	Description
Total number of destinations	The number of known destination address prefixes.
interface	The name of an interface with associated ATM tag bindings.
total	The total number of ATM tags on this interface.
active	The number of ATM tags in an "active" state that are ready to use for data transfer.
bindwait	The number of bindings that are waiting for a tag assignment from the neighbor TSR.
local	The number of ATM tags assigned by this TSR on this interface.

Continues

Table 11-6 *show tag-switching atm-tdp summary* Field Descriptions (Continued)

Field	Description
remote	The number of ATM tags assigned by the neighbor TSR on this interface.
other	The number of ATM tags in a state other than active or bindwait.
Waiting for bind on ATM0/0/0	A list of the destination address prefixes (on a particular interface) that are waiting for ATM tag assignment from the neighbor TSR.

Related Commands

You can search online at www.cisco.com to find documentation of related commands.

show tag-switching atm-tdp bindings

show tag-switching forwarding-table

To display the contents of the TFIB, use the **show tag-switching forwarding-table** privileged EXEC command.

> **show tag-switching forwarding-table** [{*network* {*mask* | *length*} | **tags** *tag* [- *tag*] | **interface** *interface* | **next-hop** *address* | **tsp-tunnel** [*tunnel-id*]}] [**detail**]

Syntax	Description
network	(Optional) Destination network number.
mask	IP address of destination mask whose entry is to be shown.
length	Number of bits in mask of destination.
tags *tag - tag*	(Optional) Shows entries with specified local tags only.
interface *interface*	(Optional) Shows entries with specified outgoing interface only.
next-hop *address*	(Optional) Shows entries with specified neighbor as next hop only.
tsp-tunnel [*tunnel-id*]	(Optional) Shows entries with specified TSP tunnel only, or all TSP tunnel entries.
detail	(Optional) Displays information in long form (includes length of encapsulation, length of MAC string, MTU, and all tags).

Command Mode

Privileged EXEC

Usage Guidelines

This command first appeared in Cisco IOS Release 11.1 CT. The optional parameters allow specification of a subset of the entire TFIB.

Sample Displays

The following is sample output from the **show tag-switching forwarding-table** command:

```
Router# show tag-switching forwarding-table

Local Outgoing      Prefix          Bytes tag Outgoing      Next Hop
tag   tag or VC     or Tunnel Id    switched  interface
26    Untagged      10.253.0.0/16   0         Et4/0/0       172.27.32.4
28    1/33          10.15.0.0/16    0         AT0/0.1       point2point
29    Pop tag       10.91.0.0/16    0         Hs5/0         point2point
      1/36          10.91.0.0/16    0         AT0/0.1       point2point
30    32            10.250.0.97/32  0         Et4/0/2       10.92.0.7
      32            10.250.0.97/32  0         Hs5/0         point2point
34    26            10.77.0.0/24    0         Et4/0/2       10.92.0.7
      26            10.77.0.0/24    0         Hs5/0         point2point
35    Untagged  [T] 10.100.100.101/32 0       Tu301         point2point
36    Pop tag       168.1.0.0/16    0         Hs5/0         point2point
      1/37          168.1.0.0/16    0         AT0/0.1       point2point

[T]    Forwarding through a TSP tunnel.
       View additional tagging info with the 'detail' option
```

The following is sample output from the **show tag-switching forwarding-table** command when you specify **detail:**

```
Router# show tag-switching forwarding-table detail

Local Outgoing      Prefix          Bytes tag Outgoing      NextHop
tag   tag or VC     or Tunnel Id    switched  interface
26    Untagged      10.253.0.0/16   0         Et4/0/0       172.27.32.4
      MAC/Encaps=0/0, MTU=1504, Tag Stack{}
28    1/33          10.15.0.0/16    0         AT0/0.1       point2point
      MAC/Encaps=4/8, MTU=4470, Tag Stack{1/33(vcd=2)}
      00020900 00002000
29    Pop tag       10.91.0.0/16    0         Hs5/0         point2point
      MAC/Encaps=4/4, MTU=4474, Tag Stack{}
      FF030081
      1/36          10.91.0.0/16    0         AT0/0.1       point2point
      MAC/Encaps=4/8, MTU=4470, Tag Stack{1/36(vcd=3)}
      00030900 00003000
30    32            10.250.0.97/32  0         Et4/0/2       10.92.0.7
      MAC/Encaps=14/18, MTU=1500, Tag Stack{32}
```

```
      006009859F2A00E0F7E984828847 00020000
   32            10.250.0.97/32    0          Hs5/0          point2point
      MAC/Encaps=4/8, MTU=4470, Tag Stack{32}
      FF030081 00020000
34 26            10.77.0.0/24      0          Et4/0/2        10.92.0.7
      MAC/Encaps=14/18, MTU=1500, Tag Stack{26}
      006009859F2A00E0F7E984828847 0001A000
   26            10.77.0.0/24      0          Hs5/0          point2point
      MAC/Encaps=4/8, MTU=4470, Tag Stack{26}
      FF030081 0001A000
35 Untagged      10.100.100.101/32 0          Tu301          point2point
      MAC/Encaps=0/0, MTU=1504, Tag Stack{}, via Et4/0/2
36 Pop tag       168.1.0.0/16      0          Hs5/0          point2point
      MAC/Encaps=4/4, MTU=4474, Tag Stack{}
      FF030081
   1/37          168.1.0.0/16      0          AT0/0.1        point2point
      MAC/Encaps=4/8, MTU=4470, Tag Stack{1/37(vcd=4)}
      00040900 00004000
```

Table 11-7 describes the significant fields in the output from the **show tag-switching forwarding-table** command.

Table 11-7 *show tag-switching forwarding-table Field Descriptions*

Field	Description
Local tag	Tag assigned by this router.
Outgoing tag or VC	Tag assigned by next hop, or VPI/VCI used to get to next hop. Some of the entries you can have in this column are
	[T] means forwarding through a TSP tunnel.
	Untagged means there is no tag for the destination from the next hop, or Tag Switching is not enabled on the outgoing interface.
	Pop tag means the next hop advertised an implicit NULL tag for the destination, and this router popped the top tag.
Prefix or Tunnel Id	Address or tunnel to which packets with this tag are going.
Bytes tag switched	Number of bytes switched with this incoming tag.
Outgoing interface	Interface through which packets with this tag are sent.
NextHop	IP address of neighbor that assigned the outgoing tag.
Mac/Encaps	Length in bytes of Layer 2 header, and length in bytes of packet encapsulation, including Layer 2 header and tag header.
MTU	MTU of tagged packet.
Tag Stack	All the outgoing tags. If the outgoing interface is TC-ATM, the VCD is also shown.
00020900 00002000	The actual encapsulation in hexadecimal form. There is a space shown between Layer 2 and tag header.

show tag-switching interfaces

To display information about one or more interfaces that have Tag Switching enabled, use the **show tag-switching interfaces** privileged EXEC command.

>**show tag-switching interfaces** [*interface*][**detail**]

Syntax	Description
interface	(Optional) The interface about which to display Tag Switching information.
detail	(Optional) Displays information in long form.

Command Mode

EXEC

Usage Guidelines

This command first appeared in Cisco IOS Release 11.1 CT. You can show information about the requested interface or about all interfaces on which Tag Switching is enabled.

Sample Display

The following is sample output from the **show tag-switching interfaces** command:

```
Router# show tag-switching interfaces

Interface          IP    Tunnel  Operational
Hssi3/0            Yes   Yes     No
ATM4/0.1           Yes   Yes     Yes        (ATM tagging)
Ethernet5/0/0      No    Yes     Yes
Ethernet5/0/1      Yes   No      Yes
Ethernet5/0/2      Yes   No      No
Ethernet5/0/3      Yes   No      Yes
Ethernet5/1/1      Yes   No      No
```

NOTE	If the interface uses TC-ATM procedures, the line in the display output is marked (ATM tagging).

Table 11-8 describes the significant fields in the output from the **show tag-switching interfaces** command.

Table 11-8 *show tag-switching interfaces* *Field Descriptions*

Field	Description
Interface	Interface name.
IP	Yes, if IP tagging has been enabled on this interface.
Tunnel	Yes, if TSP tunnel tagging has been enabled on this interface.
Operational	Operational state. Yes, if packets are being tagged.
MTU	Maximum number of data bytes per tagged packet that will be transmitted.

The following is sample output from the **show tag-switching interfaces** command when you specify **detail:**

```
Router# show tag-switching interface Ethernet2/0/1 detail

Interface Hssi3/0:
        IP tagging enabled
        TSP Tunnel tagging enabled
        Tagging not operational
        MTU = 4470
Interface ATM4/0.1:
        IP tagging enabled
        TSP Tunnel tagging enabled
        Tagging operational
        MTU = 4470
        ATM tagging: Tag VPI = 1, Control VC = 0/32
Interface Ethernet5/0/0:
        IP tagging not enabled
        TSP Tunnel tagging enabled
        Tagging operational
        MTU = 1500
Interface Ethernet5/0/1:
        IP tagging enabled
        TSP Tunnel tagging not enabled
        Tagging operational
        MTU = 1500
```

```
Interface Ethernet5/0/2:
      IP tagging enabled
      TSP Tunnel tagging not enabled
      Tagging not operational
      MTU = 1500
Interface Ethernet5/0/3:
      IP tagging enabled
      TSP Tunnel tagging not enabled
      Tagging operational
      MTU = 1500
```

Related Commands

You can search online at www.cisco.com to find documentation of related commands.

tag-switching ip (interface configuration)
tag-switching tsp-tunnels (interface configuration)

show tag-switching tdp bindings

To display the contents of the TIB, use the **show tag-switching tdp bindings** privileged EXEC command.

show tag-switching tdp bindings [*network* {*mask* | *length*} [**longer-prefixes**]]
[**local-tag** *tag* [- *tag*]} **remote-tag** *tag* [- *tag*] [**neighbor** *address*] [**local**]

Syntax	Description
network	(Optional) Destination network number.
mask	Network mask written as A.B.C.D.
length	Mask length (1 to 32 characters).
longer-prefixes	(Optional) Selects any prefix that matches *mask* with *length* to 32.
local-tag *tag* - *tag*	(Optional) Displays entries matching local tag values by this router. Use the - *tag* argument to indicate tag range.
remote-tag *tag* - *tag*	(Optional) Displays entries matching tag values assigned by a neighbor router. Use the - *tag* argument to indicate tag range.
neighbor *address*	(Optional) Displays tag bindings assigned by selected neighbor.
local	(Optional) Displays local tag bindings.

Command Mode

Privileged EXEC

Usage Guidelines

This command first appeared in Cisco IOS Release 11.1 CT.

A request can specify that the entire database be shown, or it can be limited to a subset of entries. A request to show a subset of entries can be based on the prefix, on input or output tag values or on ranges, and/or the neighbor advertising the tag.

Sample Displays

The following is sample output from the **show tag-switching tdp bindings** command. This form of the command causes the contents of the entire TIB to be displayed.

```
Router# show tag-switching tdp bindings

Matching entries:
  tib entry: 10.92.0.0/16, rev 28
       local binding:  tag: imp-null(1)
       remote binding: tsr: 172.27.32.29:0, tag: imp-null(1)
  tib entry: 10.102.0.0/16, rev 29
       local binding:  tag: 26
       remote binding: tsr: 172.27.32.29:0, tag: 26
  tib entry: 10.105.0.0/16, rev 30
       local binding:  tag: imp-null(1)
       remote binding: tsr: 172.27.32.29:0, tag: imp-null(1)
  tib entry: 10.205.0.0/16, rev 31
       local binding:  tag: imp-null(1)
       remote binding: tsr: 172.27.32.29:0, tag: imp-null(1)
  tib entry: 10.211.0.7/32, rev 32
       local binding:  tag: 27
       remote binding: tsr: 172.27.32.29:0, tag: 28
  tib entry: 10.220.0.7/32, rev 33
       local binding:  tag: 28
       remote binding: tsr: 172.27.32.29:0, tag: 29
  tib entry: 99.101.0.0/16, rev 35
       local binding:  tag: imp-null(1)
       remote binding: tsr: 172.27.32.29:0, tag: imp-null(1)
  tib entry: 100.101.0.0/16, rev 36
       local binding:  tag: 29
       remote binding: tsr: 172.27.32.29:0, tag: imp-null(1)
  tib entry: 171.69.204.0/24, rev 37
       local binding:  tag: imp-null(1)
       remote binding: tsr: 172.27.32.29:0, tag: imp-null(1)
  tib entry: 172.27.32.0/22, rev 38
       local binding:  tag: imp-null(1)
       remote binding: tsr: 172.27.32.29:0, tag: imp-null(1)
  tib entry: 210.10.0.0/16, rev 39
       local binding:  tag: imp-null(1)
```

```
tib entry: 210.10.0.8/32, rev 40
      remote binding: tsr: 172.27.32.29:0, tag: 27
```

The following is sample output from the **show tag tdp bindings 10.0.0.0 8 longer-prefixes neighbor 172.27.32.29** variant of the command; it displays tags learned from TSR 172.27.32.29 for network 10.0.0.0 and any of its subnets. The use of the **neighbor** option suppresses the output of local tags and tags learned from other neighbors.

```
Router# show tag tdp bindings 10.0.0.0 8 longer-prefixes neighbor 172.27.32.29

tib entry: 10.92.0.0/16, rev 28
      remote binding: tsr: 172.27.32.29:0, tag: imp-null(1)
tib entry: 10.102.0.0/16, rev 29
      remote binding: tsr: 172.27.32.29:0, tag: 26
tib entry: 10.105.0.0/16, rev 30
      remote binding: tsr: 172.27.32.29:0, tag: imp-null(1)
tib entry: 10.205.0.0/16, rev 31
      remote binding: tsr: 172.27.32.29:0, tag: imp-null(1)
tib entry: 10.211.0.7/32, rev 32
      remote binding: tsr: 172.27.32.29:0, tag: 28
tib entry: 10.220.0.7/32, rev 33
      remote binding: tsr: 172.27.32.29:0, tag: 29
```

Table 11-9 describes the significant fields in the **show tag-switching tdp bindings** display.

Table 11-9   *show tag-switching tdp bindings Field Descriptions*

Field	Description
tib entry	Indicates that the following lines are the TIB entry for a particular destination (network/mask). The revision number is used internally to manage tag distribution for this destination.
remote binding	A list of outgoing tags for this destination learned from other TSRs. Each item on this list identifies the TSR from which the outgoing tag was learned and the tag itself. The TSR is identified by its TDP identifier.
imp-null	The implicit null tag. This tag value instructs the upstream router to pop the tag entry off the tag stack before forwarding the packet.

Related Commands

You can search online at www.cisco.com to find documentation of related commands.

show tag-switching forwarding-table
show tag-switching tdp neighbors

show tag-switching tdp discovery

To display the status of the TDP discovery process, use the **show tag-switching tdp discovery** privileged EXEC command. Status means a list of interfaces over which TDP discovery is running.

>   **show tag-switching tdp discovery**

Syntax Description

This command has no arguments or keywords.

Command Mode

Privileged EXEC

Usage Guidelines

This command first appeared in Cisco IOS Release 11.1 CT.

Sample Display

The following is sample output from the **show tag-switching tdp discovery** command:

```
Router# show tag-switching tdp discovery

Local TDP Identifier:
    172.27.32.29:0
TDP Discovery Sources:
    Interfaces:
ATM0/0.1:       xmit/recv
ATM0/0.1:       xmit/rec
Ethernet4/0/1:  xmit/recv
Ethernet4/0/2:  xmit/recv
POS6/0/0:       xmit/recv
```

Table 11-10 describes the significant fields in the output from the **show tag-switching tdp discovery** command.

Table 11-10 *show tag-switching tdp discovery Field Descriptions*

Field	Description
Local TDP Identifier	The TDP identifier for the local router. A TDP identifier is a 6-byte quantity displayed as an IP address:number.
	The Cisco convention is to use a router ID for the first 4 bytes of the TDP identifier, and integers starting with 0 for the final two bytes of the IP address:number.

Continues

Table 11-10 *show tag-switching tdp discovery* Field Descriptions (Continued)

Field	Description
Interfaces	Lists the interfaces engaging in TDP discovery activity. xmit indicates that the interface is transmitting TDP discovery hello packets; recv indicates that the interface is receiving TDP discovery hello packets.

Related Commands

You can search online at www.cisco.com to find documentation of related commands.

show tag-switching tdp neighbors

show tag-switching tdp neighbors

To display the status of TDP sessions, enter the **show tag-switching tdp neighbor** privileged EXEC command.

show tag-switching tdp neighbors [*address* | *interface*] [**detail**]

Syntax	Description
address	(Optional) The neighbor with this IP address.
interface	(Optional) TDP neighbors accessible over this interface.
detail	(Optional) Displays information in long form.

Command Mode

Privileged EXEC

Usage Guidelines

This command first appeared in Cisco IOS Release 11.1 CT.

The neighbor information branch can give information about all TDP neighbors, or it can be limited to

● The neighbor with a specific IP address

● TDP neighbors known to be accessible over a specific interface

Sample Display

The following is sample output from the **show tag-switching tdp neighbors** command:

```
Router# show tag-switching tdp neighbors

Peer TDP Ident: 10.220.0.7:1; Local TDP Ident 172.27.32.29:1
        TCP connection: 10.220.0.7.711 - 172.27.32.29.11029
        State: Oper; PIEs sent/rcvd: 17477/17487; Downstream on demand
Up time: 01:03:00
TDP discovery sources:
        ATM0/0.1
Peer TDP Ident: 210.10.0.8:0; Local TDP Ident 172.27.32.29:0
        TCP connection: 210.10.0.8.11004 - 172.27.32.29.711
        State: Oper; PIEs sent/rcvd: 14656/14675; Downstream
Up time: 2d5h
        TDP discovery sources:
        Ethernet4/0/1
        Ethernet4/0/2
        POS6/0/0
        Addresses bound to peer TDP Ident:
        99.101.0.8      172.27.32.28    10.105.0.8      10.92.0.8
        10.205.0.8      210.10.0.8
```

Table 11-11 describes the significant fields in the output from the **show tag-switching tdp neighbors** command.

Table 11-11   *show tag-switching tdp neighbors Field Descriptions*

Field	Description
Peer TDP Ident	The TDP identifier of the neighbor (peer device) for this session.
Local TDP Ident	The TDP identifier for the local TSR for this session.
TCP connection	The TCP connection used to support the TDP session. The format for displaying the TCP connection is *peer IP address.peer port* *local IP address.local port*
State	The state of the TDP session. Generally this is Oper (operational), but transient is another possible state.
PIEs sent/rcvd	The number of TDP protocol information elements (PIEs) sent to and received from the session peer device. The count includes the transmission and receipt of periodic keep alive PIEs, which are required for maintenance of the TDP session.
Downstream	Indicates that the downstream method of tag distribution is being used for this TDP session. When the downstream method is used, a TSR advertises all of its locally assigned (incoming) tags to its TDP peer device (subject to any configured access list restrictions).

Continues

Table 11-11 *show tag-switching tdp neighbors Field Descriptions (Continued)*

Field	Description
Downstream on demand	Indicates that the downstream on demand method of tag distribution is being used for this TDP session. When the downstream on demand method is used, a TSR advertises its locally assigned (incoming) tags to its TDP peer device only when the peer device asks for them.
Up time	The length of time the TDP session has existed.
TDP discovery sources	The source(s) of TDP discovery activity that led to the establishment of this TDP session.
Addresses bound to peer TDP Ident	The known interface addresses of the TDP session peer device. These are addresses that may appear as next-hop addresses in the local routing table. They are used to maintain the TFIB.

Related Commands

You can search online at www.cisco.com to find documentation of related commands.

show tag-switching tdp discovery

show tag-switching tdp parameters

To display available TDP parameters, use the **show tag-switching tdp parameters** privileged EXEC command.

> **show tag-switching tdp parameters**

Syntax Description

This command has no arguments or keywords.

Command Mode

Privileged EXEC

Usage Guidelines

This command first appeared in Cisco IOS Release 11.1 CT.

Sample Display

The following is sample output from the **show tag-switching tdp parameters** command:

```
Router# show tag-switching tdp parameters

Protocol version: 1
 Downstream tag pool: min tag: 10; max_tag: 10000; reserved tags: 16
 Session hold time: 15 sec; keep alive interval: 5 sec
 Discovery hello: holdtime: 15 sec; interval: 5 sec
 Discovery directed hello: holdtime: 15 sec; interval: 5 sec
 Accepting directed hellos
```

Table 11-12 describes the significant fields in the output from the **show tag-switching tdp parameters** command.

Table 11-12   *show tag-switching tdp parameters Command Field Descriptions*

Field	Description
Protocol version	Indicates the version of the TDP running on the platform.
Downstream tag pool	Describes the range of tags available for the platform to assign for Tag Switching. The tags available run from the smallest tag value (min tag) to the largest tag value (max tag), with a modest number of tags at the low end of the range (reserved tags) reserved for diagnostic purposes.
Session hold time	Indicates the time to maintain a TDP session with a TDP peer device without receiving TDP traffic or a TDP keepalive from the peer device.
keep alive interval	Indicates the interval of time between consecutive transmission TDP keep alive messages to a TDP peer device.
Discovery hello	Indicates the amount of time to remember that a neighbor platform wants a TDP session without receiving a TDP hello from the neighbor (holdtime), and the time interval between transmitting TDP Hello messages to neighbors (interval).
Discovery directed hello	Indicates the amount of time to remember that a neighbor platform wants a TDP session when (1) the neighbor platform is not directly connected to the router and (2) the neighbor platform has not sent a TDP hello message. The interval is known as hold time.
	Also indicates the time interval between the transmission of Hello messages to a neighbor not directly connected to the router.
Accepting directed hellos	Indicates that the platform will accept and act on directed TDP hello messages (may not be present).

Related Commands

You can search online at www.cisco.com to find documentation of related commands.

tag-switching tdp discovery
tag-switching tdp holdtime

show tag-switching tsp-tunnels

To display information about the configuration and status of selected tunnels, use the **show tag-switching tsp-tunnels** privileged EXEC command.

> **show tag-switching tsp-tunnels** [{**head** | **middle** | **tail** | **all** | **remote** | *address*}
> [*interface-number*]] [**brief**]

Syntax	Description
head	(Optional) Displays information for tunnels that originate at the node.
middle	(Optional) Displays information for tunnels that pass through the node.
tail	(Optional) Displays information for tunnels that terminate at the node.
all	(Optional) Displays the combination of head, middle, and tail information for tunnels.
remote	(Optional) Displays information for tunnels that originate elsewhere; thus, it is the combination of middle and tail.
address	(Optional) Displays information for tunnels that use the specified address in their identifier.
interface-number	(Optional) Displays information for tunnels that use the specified number in their identifier.
brief	(Optional) Displays a brief summary of tunnel status and configuration.

Command Mode

Privileged EXEC

Usage Guidelines

This command first appeared in Cisco IOS Release 11.1 CT.

The optional keywords restrict the set of tunnels displayed. With no optional keywords, the command displays all tunnels passing through the node.

Each TSP tunnel has a globally unique identifier. When signalling the TSP tunnel is signalled and is available at each hop, this identifier is used. This identifier is a combination of the originating IP address and the number of the Cisco IOS tunnel interface used in configuring the TSP tunnel at the headend.

Sample Display

The following is sample output from the **show tag-switching tsp-tunnels** command:

```
Signalling Summary:
            TSP Tunnels Process:            running
            RSVP Process:                   running
            Forwarding:                     enabled
TUNNEL ID                 DESTINATION     STATUS          CONNECTION
10.106.0.6 0              10.2.0.12       up              up
```

Table 11-13 describes the significant fields in the output from the **show tag-switching tsp-tunnels** command.

Table 11-13 *show tag-switching tsp-tunnels* Field Descriptions

Field	Description
Signalling Summary	The status of the signalling and forwarding mechanism that is required in order for TSP tunnels to be signalled through the router.
TSP Tunnels Process	The status of the TSP tunnel signalling process. This process interacts with the signalling protocol to manage signalled tunnels and monitors the state of established tunnels.
RSVP Process	The status of the RSVP process. You use the RSVP protocol to signal tunnels.
Forwarding	The status of the forwarding mechanism used to switch data through local TSP tunnel segments.
TUNNEL ID	The identity of the tunnel being summarized as shown in the previous display output. The tunnel ID includes an IP address part and a number part, and is unique within the entire network.
DESTINATION	The destination of the TSP tunnel being summarized as shown in the previous display output—the IP address of the tunnel tail.
STATUS	The configuration status of the tunnel. At the head, this is an indication of whether or not the tunnel has been completely configured. It also refers to the status of the associated software and hardware interfaces.

Continues

Table 11-13 *show tag-switching tsp-tunnels Field Descriptions (Continued)*

Field	Description
CONNECTION	The connection status of the tunnel. This is an indication of whether or not the local signalling/configuration information shows that the tunnel is up. Typically the tunnel becomes up at the tail hop first, at the second to the last hop, and so forth until signalling brings it up at the first hop.

Related Commands

You can search online at www.cisco.com to find documentation of related commands.

tag-switching tsp-tunnels (interface configuration)
tunnel mode tag-switching

tag-switching advertise-tags

To control the distribution of locally assigned (incoming) tags via the TDP, use the **tag-switching advertise-tags** global configuration command. To disable tag advertisement, use the **no** form of this command.

> **tag-switching advertise-tags** [**for** *access-list-number* [**to** *access-list-number*]
> **no tag-switching advertise-tags** [**for** *access-list-number* [**to** *access-list-number*]

Syntax	Description
for *access-list-number*	(Optional) Specifies which destinations should have their tags advertised.
to *access-list-number*	(Optional) Specifies which TSR neighbors should receive tag advertisements. A TSR is identified by the router ID that is the first 4 bytes of its 6-byte TDP identifier.

Default

Advertise all to all is the default.

Command Mode

Global configuration

Usage Guidelines

This command first appeared in Cisco IOS Release 11.1 CT.

To enable the distribution of all locally assigned tags to all TDP neighbors, use the **tag-switching advertise-tags** command.

You can enter multiple **tag-switching advertise-tags** commands. Taken together, they determine how local tags are advertised.

NOTE   This command has no effect for a TC-ATM interface. The effect is always as if the **tag-switching advertise-tags** command had been executed.

Examples

In the following example, the router is configured to advertise all locally assigned tags to all TDP neighbors. This is the default:

```
Router(config)# tag-switching advertise-tags
```

In the following example, the router is configured to advertise to all TDP neighbors tags for networks 10.101.0.0 and 10.221.0.0 only:

```
Router(config)# access-list 1 permit 10.101.0.0 0.0.255.255
Router(config)# access-list 4 permit 10.221.0.0 0.0.255.255
Router(config)# tag-switching advertise-tags for 1
Router(config)# tag-switching advertise-tags for 4
```

In the following example, the router is configured to advertise all tags to all TDP neighbors except neighbor 10.101.0.8:

```
Router(config)# access-list 1 permit any
Router(config)# access-list 2 deny  10.101.0.8
Router(config)# tag-switching advertise-tags
Router(config)# tag-switching advertise-tags for 1 to 2
```

tag-switching atm allocation-mode

To control the mode used for handling tag binding requests on TC-ATM interfaces, use the **tag-switching atm allocation-mode** global configuration command. Use the **no** form of this command to disable this feature.

> **tag-switching atm allocation-mode {optimistic | conservative}**
> **no tag-switching atm allocation-mode {optimistic | conservative}**

Syntax	Description
optimistic	Tag binding is returned immediately and packets are discarded until the downstream setup is complete.
conservative	Tag binding is delayed until the tag VC has been set up downstream.

Default

The default is conservative.

Command Mode

Global configuration

Usage Guidelines

This command first appeared in Cisco IOS Release 11.1 CT.

Example

In the following example, the mode for handling binding requests is set to optimistic on a TC-ATM interface:

```
tag-switching atm allocation-mode optimistic
```

tag-switching atm control-vc

To configure the VPI and VCI to be used for the initial link to the Tag Switching peer device, use the **tag-switching atm control-vc** interface configuration command. The initial link is used to establish the TDP session and to carry non-IP traffic. To clear the interface configuration, use the **no** form of this command.

> **tag-switching atm control-vc** *vpi vci*
> **no tag-switching atm control-vc** *vpi vci*

Syntax	Description
vpi	Virtual path identifier.
vci	Virtual channel identifier.

Default

If the subinterface has not changed to a VP tunnel, the default is 0/32. If the subinterface corresponds to VP tunnel VPI X, the default is X/32.

Command Mode

Interface configuration

Usage Guidelines

This command first appeared in Cisco IOS Release 11.1 CT.

For a router interface (for example, an AIP) ATM Tag Switching can be enabled only on a tag-switch subinterface.

NOTE   The **tag-switching atm control-vc and tag-switching atm vpi** subinterface level configuration commands are available on any interface that can support ATM tagging.

On the Cisco LightStream 1010 ATM switch, a subinterface corresponds to a VP tunnel, so the VPI field of the control-vc must match the VPI field of the VP tunnel.

Example

The following creates a Tag Switching subinterface on a router and selects VPI 1 and VCI 34 as the control VC:

```
interface atm4/0.1 tag-switching
 tag-switching ip
 tag-switching atm control-vc 1 34
```

Related Commands

You can search online at www.cisco.com to find documentation of related commands.

show tag-switching atm-tdp capability
show tag-switching interfaces

tag-switching atm maxhops

To limit the maximum hop counts to a value you have specified, use the **tag-switching atm maxhops** global configuration command. Use the **no** form of this command to ignore the hop count.

> **tag-switching atm maxhops** [*number*]
> **no tag-switching atm maxhops**

Syntax

number

Description

(Optional) Maximum hop count.

Default

The default is 254.

Command Mode

Global configuration

Usage Guidelines

This command first appeared in Cisco IOS Release 11.1 CT.

When an ATM TSR receives a BIND REQUEST, it does not send a BIND back if the value in the request is equal to the maxhops value. Instead, the ATM TSR or TSR returns an error that specifies that the hop count has been reached.

When an ATM TSR initiates a request for a tag binding, it includes a parameter specifying the maximum number of hops that the request should travel before reaching the edge of the ATM Tag Switching region. This is used to prevent forwarding loops from setting up tag paths across the ATM region.

Example

The following example sets the hop count limit to 2:

```
tag-switching atm maxhops 2
```

Related Commands

You can search online at www.cisco.com to find documentation of related commands.

show tag-switching atm-tdp bindings

tag-switching atm vc-merge

To control whether vc-merge (multipoint-to-point) is supported for unicast tag VCs, use the **tag-switching atm vc-merge** global configuration command. Use the **no** form of this command to disable this feature.

> **tag-switching atm vc-merge**
> **no tag-switching atm vc-merge**

Syntax Description

This command has no arguments or keywords.

Default

The default is enabled if the hardware supports the ATM-VC merge capability.

Command Mode

Global configuration

Usage Guidelines

This command first appeared in Cisco IOS Release 11.1 CT.

Example

The following example disables VC merge:

```
no tag-switching atm vc-merge
```

Related Commands

You can search online at www.cisco.com to find documentation of related commands.

show tag-switching atm-tdp capability

tag-switching atm vpi

To configure the range of values to use in the VPI field for tag VCs, use the **tag-switching atm vpi** interface configuration command. To clear the interface configuration, use the **no** form of this command.

> **tag-switching atm vpi** *vpi* [- *vpi*]
> **no tag-switching atm vpi** *vpi* [- *vpi*]

Syntax	Description
vpi	Virtual path identifier (low end of range).
- vpi	(Optional) Virtual path identifier (high end of range).

Default

The default is 1-1.

Command Mode

Interface configuration

Usage Guidelines

This command first appeared in Cisco IOS Release 11.1 CT.

To configure ATM Tag Switching on a router interface (for example, an ATM interface processor), you must enable a Tag Switching subinterface.

NOTE   The **tag-switching atm control-vc** and **tag-switching atm vpi** interface configuration commands are available on any interface that can support ATM tagging.

Use this command to select an alternate range of VPI values for ATM tag assignment on this interface. The two ends of the link negotiate a range defined by the intersection of the range configured at each end.

Example

The following example creates a subinterface and selects a VPI range from VPI 1 to VPI 3:

```
interface atm4/0.1 tag-switching
 tag-switching ip
 tag-switching atm vpi 1-3
```

Related Commands

You can search online at www.cisco.com to find documentation of related commands.

tag-switching atm control-vc

tag-switching ip (global configuration)

To allow Tag Switching of IPv4 packets, use the **tag-switching ip** global configuration command. To disable IP Tag Switching across all interfaces, use the **no** form of this command.

> **tag-switching ip**
> **no tag-switching ip**

Syntax Description

This command has no arguments or keywords.

Default

Tag Switching of IPv4 packets is allowed.

Command Mode

Global configuration

Usage Guidelines

This command first appeared in Cisco IOS Release 11.1 CT.

Dynamic Tag Switching (that is, distribution of tags based on routing protocols) is allowed by this optional command, but it is not actually enabled until the interface-level **tag-switching ip** command is issued on at least one interface. The **no** form of this command stops the distribution of dynamic tags and the sending of outgoing tagged packets on all interfaces. The command does not affect the sending of tagged packets through TSP tunnels.

For a TC-ATM interface, the **no** form of this command prevents the establishment of tag VCs beginning at, terminating at, or passing through the platform.

Example

The following example prevents the distribution of dynamic tags on all interfaces:

```
configure terminal
no tag-switching ip
```

Related Commands

You can search online at www.cisco.com to find documentation of related commands.

tag-switching ip (interface configuration)

tag-switching ip (interface configuration)

To enable Tag Switching of IPv4 packets on an interface, use the **tag-switching ip** interface configuration command. To disable IP Tag Switching on this interface, use the **no** form of this command.

> **tag-switching ip**
> **no tag-switching ip**

Syntax Description

This command has no arguments or keywords.

Default

Tag Switching of IPv4 packets is disabled on this interface.

Command Mode

Interface configuration

Usage Guidelines

This command first appeared in Cisco IOS Release 11.1 CT.

The first time this command is issued on any interface, dynamic Tag Switching is enabled on the router as a whole. TDP hello messages are issued on this interface. When an outgoing tag for a destination routed out through this interface is received, packets sent to that destination are assigned with that tag.

The **no** form of this command causes packets routed out through this interface to be sent untagged, and outgoing TDP hello messages are no longer sent.

When the **no** form is issued on the only interface of a router for which Tag Switching was enabled, dynamic Tag Switching is disabled on the router as a whole.

For a TC-ATM interface, the **no** form of this command prevents the establishment of tag VCs beginning at, terminating at, or passing through the platform.

Example

The following example enables Tag Switching on the specified Ethernet interface:

```
configure terminal
interface e0/2
  tag-switching ip
```

Related Commands

You can search online at www.cisco.com to find documentation of related commands.

tag-switching advertise-tags
show tag-switching interfaces

tag-switching mtu

To override the per-interface MTU, use the **tag-switching mtu** interface configuration command. To restore the default, use the **no** form of this command.

> **tag-switching mtu** *bytes*
> **no tag-switching mtu**

Syntax   Description

bytes   MTU in bytes.

Default

Minimum is 128 bytes; maximum depends on interface medium type.

Command Mode

Interface configuration

Usage Guidelines

This command first appeared in Cisco IOS Release 11.1 CT.

If a tagged IP packet exceeds the MTU set for the interface, the Cisco IOS software will fragment it. All devices on a physical medium must have the same protocol MTU in order to operate.

NOTE   Changing the MTU value (with the **mtu** interface configuration command) can affect the tag IP MTU value. If the current tag IP MTU value is the same as the MTU value and you change the MTU value, the tag IP MTU value will be modified automatically to match the new MTU. However, the reverse is not true; changing the tag IP MTU value has no effect on the value for the **mtu** command.

Example

The following example sets the maximum tagged packet size for the first serial interface to 300 bytes:

```
interface serial 0
 tag-switching mtu 300
```

tag-switching tag-range downstream

To configure the size of the tag space for downstream unicast tag allocation, use the **tag-switching tag-range downstream** global configuration command. Use the **no** form of this command to revert to the platform defaults.

> **tag-switching tag-range downstream** *min max reserved*
> **no tag-switching tag-range downstream** *min max reserved*

Syntax	Description
min	The smallest tag allowed in the tag space. The default is 10.
max	The largest tag allowed in the tag space. The default is 10,000.
reserved	The number of tags reserved for diagnostic purposes. These tags come out of the low end of the tag space. The default is 16.

Default

The default values for the parameters are as follows:

min—10
max—10,000
reserved—16

Command Mode

Global configuration

Usage Guidelines

This command first appeared in Cisco IOS Release 11.1 CT.

Example

The following example shows how to configure the size of the tag space for downstream unicast tag allocation. In the example, *min* is set with the value of 10, *max* is set with the value of 12000, and *reserved* is set with the value of 16.

```
configure terminal
tag-switching tag-range downstream 10 12000 16
```

Related Commands

You can search online at www.cisco.com to find documentation of related commands.

show tag-switching tdp parameters

tag-switching tdp discovery

To configure the interval between transmission of TDP discovery hello messages or the hold time for a TDP transport connection, use the **tag-switching tdp discovery** global configuration command.

tag-switching tdp discovery {**hello | directed hello**} {**holdtime | interval**} *seconds*

Syntax	Description
hello	Configures the intervals and hold times for directly connected neighbors.

directed-hello	Configures the intervals and hold times for neighbors that are not directly connected (for example, TDP sessions that run through a TSP tunnel).
holdtime	The interval for which a connection stays up if no hello messages are received. The default is 15 seconds.
interval	The period between the sending of consecutive hello messages. The default is 5 seconds.
seconds	The hold time or interval.

Default

The default values for **holdtime** and **interval** are

> **holdtime**—15 seconds
> **interval**—5 seconds

Command Mode

Global configuration

Usage Guidelines

This command first appeared in Cisco IOS Release 11.1 CT.

Example

In the following example, the interval for which a connection stays up if no hello messages are received is set to 5 seconds:

```
tag-switching tdp discovery hello holdtime 5
```

Related Commands

You can search online at www.cisco.com to find documentation of related commands.

show tag-switching tdp parameters
tag-switching tdp holdtime

tag-switching tdp holdtime

To enable TSP tunnel functionality on a device, use the **tag-switching tdp holdtime** global configuration command.

tag-switching tdp holdtime *seconds*

Syntax	Description
seconds	The time for which a TDP session is maintained in the absence of TDP messages from the session peer device.

Default

15 seconds

Command Mode

Global configuration

Usage Guidelines

This command first appeared in Cisco IOS Release 11.1 CT.

When a TDP session is initiated, the hold time is set to the lower of the values configured at the two ends.

Example

In the following example, the hold time of TDP sessions is configured for 30 seconds:

```
tag-switching tdp holdtime 30
```

Related Commands

You can search online at www.cisco.com to find documentation of related commands.

show tag-switching tdp parameters
tag-switching tdp discovery

tag-switching tsp-tunnels (global configuration)

To allow the operation of TSP tunnels, use the **tag-switching tsp-tunnels** global configuration command. To disable the operation of TSP tunnels, use the **no** form of this command.

> **tag-switching tsp-tunnels**
> **no tag-switching tsp-tunnels**

Syntax Description

This command has no arguments or keywords.

Default

Disabled

Command Mode

Global configuration

Usage Guidelines

This command first appeared in Cisco IOS Release 11.1 CT.

TSP tunnel operation is allowed on the device by this optional command, but proper operation also requires that the interface-level **tag-switching tsp-tunnels** command be issued on the interfaces that are used by TSP tunnels. The **no** form of this command completely disables TSP tunnel operation on the device.

Example

The following example allows TSP tunnel operation on a device:

```
configure terminal
ip cef distributed
tag-switching tsp-tunnels
```

Related Commands

You can search online at www.cisco.com to find documentation of related commands.

ip cef distributed
show tag-switching tsp-tunnels

tag-switching tsp-tunnels (interface configuration)

To allow TSP tunnel operation over an interface, use the **tag-switching tsp-tunnels** interface configuration command. To disable TSP tunnel operation over an interface, use the **no** form of this command.

> **tag-switching tsp-tunnels**
> **no tag-switching tsp-tunnels**

Syntax Description

This command has no arguments or keywords.

Default

Disabled

Command Mode

Interface configuration

Usage Guidelines

This command first appeared in Cisco IOS Release 11.1 CT.

TSP tunnel operation over a specific interface is allowed by this optional command. In order for TSP tunnels to operate over an interface, the **tag-switching tsp-tunnels** global configuration command must also be enabled. The **no** form of this command disables TSP tunnel operation over the specified interface.

Example

The following example allows TSP tunnel operation over an interface:

```
configure terminal
ip cef distributed
tag-switching tsp-tunnels
```

Related Commands

You can search online at www.cisco.com to find documentation of related commands.

ip cef distributed
show tag-switching tsp-tunnels

traffic-engineering filter

To specify a filter with the given number and properties, use the **traffic-engineering filter** command. To disable this function, use the **no** form of this command.

traffic-engineering filter *filter-number* **egress** *ip-address mask*
no traffic-engineering filter

Syntax

filter-number

egress *ip-address mask*

Description

A decimal value representing the number of the filter.

IP address and mask for the egress port.

Command Mode

Router configuration

Usage Guidelines

This command first appeared in Cisco IOS Release 11.1 CT.

You must specify that the egress is the indicated address/mask, where egress is either the destination or the BGP next hop.

Example

In the following example, configuration is provided for the traffic engineering routing process, a traffic engineering filter, and a traffic engineering route for that filter over a TSP-encapsulated tunnel:

```
router traffic-engineering
traffic-engineering filter 5 egress 83.0.0.1 255.255.255.255
traffic-engineering route 5 tunnel 5
```

Related Commands

You can search online at www.cisco.com to find documentation of related commands.

show ip traffic-engineering routes
traffic-engineering route

traffic-engineering route

To configure a route for a specified filter, through a specified tunnel, use the **traffic-engineering route** command. To disable this function, use the **no** form of this command.

> **traffic-engineering route** *filter-number interface* [**preference** *number*] [**loop-prevention** {**on** | **off**}]

> **no traffic-engineering route** *filter-number interface* [**preference** *number*] [**loop-prevention** {**on** | **off**}]

Syntax	Description
filter-number	The number of the traffic engineering filter to be forwarded through the use of this traffic engineering route, if the route is installed.
interface	TSP-encapsulated tunnel on which traffic-passing filter should be sent, if this traffic engineering route is installed.
preference *number*	(Optional) This is a number between 1 and 255, with a lower value being more desirable. The default is 1.
loop-prevention	(Optional) This can be on or off. The default is on.

Defaults

The default values for the following parameters are

> **preference**—1
> **loop-prevention**—on

Command Mode

Router configuration

Usage Guidelines

This command first appeared in Cisco IOS Release 11.1 CT.

The traffic engineering process is used to decide if a configured traffic engineering route should be installed in the forwarding table.

The first step is to determine if the route is up. If the route is enabled, the TSP tunnel interface is up, the loop prevention check is either disabled or passed, and the traffic engineering route is up.

If multiple routes for the same filter are up, a route is selected based on administrative preference.

If loop prevention is enabled, metrics are solicited from the tunnel tail, and the loop prevention algorithm is run on the result. For a discussion of the loop prevention algorithm, see the **show ip traffic-engineering metrics** command.

Example

In the following example, configuration is provided for the traffic engineering routing process, a traffic engineering filter, and a traffic engineering route for that filter through a TSP-encapsulated tunnel:

```
router traffic-engineering
traffic-engineering filter 5 egress 83.0.0.1 255.255.255.255
traffic-engineering route 5 tunnel 5
```

Related Commands

You can search online at www.cisco.com to find documentation of related commands.

show ip traffic-engineering configuration
show ip traffic-engineering routes

tunnel mode tag-switching

To set the encapsulation mode of the tunnel to Tag Switching, use the **tunnel mode tag-switching** interface configuration command. Use the **no** form of this command to set the tunneling encapsulation mode to the default, Generic Routing Encapsulation (GRE).

tunnel mode tag-switching
no tunnel mode tag-switching

Syntax Description

This command has no arguments or keywords.

Command Mode

Interface configuration

Usage Guidelines

This command first appeared in Cisco IOS Release 11.1 CT.

A tunnel interface number must be less than or equal to 65535.

The **tunnel mode tag-switching** command fails if the interface number is invalid for a TSP tunnel identifier.

Example

In the following example, the tunnel mode is set to Tag Switching:

```
interface tunnel 5
 tunnel mode tag-switching
```

Related Commands

You can search online at www.cisco.com to find documentation of related commands.

interface tunnel
tunnel tsp-hop

tunnel tsp-hop

To define hops in the path for the Tag Switching tunnel, use the **tunnel tsp-hop** interface configuration command. Use the **no** form of this command to remove these hops.

> **tunnel tsp-hop** *hop-number ip-address* [**lasthop**]
> **no tunnel tsp-hop** *hop-number ip-address* [**lasthop**]

Syntax	Description
hop-number	The sequence number of the hop being defined in the path. The first number is 1, which identifies the hop just after the head hop.
ip-address	The IP address of the input interface on that hop.
lasthop	(Optional) Indicates that the hop being defined is the final hop in the path (the tunnel destination).

Default

No hops are defined.

Command Mode

Interface configuration

Usage Guidelines

This command first appeared in Cisco IOS Release 11.1 CT.

The list of tunnel hops must specify a strict source route for the tunnel. In other words, the router at hop <N> must be directly connected to the router at hop <N>+1.

Example

The following example shows the configuration of a two-hop tunnel. The first hop router/switch is 82.0.0.2, and the second and last hop is router/switch 81.0.0.2.

```
interface tunnel 5
 tunnel mode tag-switching
 ip unnumbered e0/1
 tunnel tsp-hop 1 82.0.0.2
 tunnel tsp-hop 2 81.0.0.2 lasthop
```

Related Commands

You can search online at www.cisco.com to find documentation of related commands.

interface tunnel
tunnel mode tag-switching

PART V

Multilayer Switching

CHAPTER 12

Multilayer Switching Overview

This chapter provides an overview of Multilayer Switching (MLS).

NOTE   The commands and configurations described in this chapter apply only to the devices that provide routing services. Commands and configurations for Catalyst 5000 series switches are not documented here.

MLS provides high-performance Layer 3 switching for the Catalyst 5000 series LAN switches. MLS switches IP data packets between subnets using advanced application-specific integrated circuit (ASIC) switching hardware. Standard routing protocols, such as Open Shortest Path First (OSPF), Enhanced Interior Gateway Routing Protocol (EIGRP), Routing Information Protocol (RIP), and Intermediate System-to-Intermediate System (IS-IS), are used for route determination.

MLS enables hardware-based Layer 3 switching to offload routers from forwarding unicast IP data packets over shared-media networking technologies such as Ethernet. The packet forwarding function is moved onto Layer 3 Catalyst 5000 series switches whenever a partial or complete switched path exists between two hosts. Packets that do not have a partial or complete switched path to reach their destinations still use routers for forwarding packets.

MLS also provides traffic statistics as part of its switching function. These statistics are used for identifying traffic characteristics for administration, planning, and troubleshooting. MLS uses NetFlow Data Export (NDE) to export the flow statistics.

The Route Switch Module (RSM) performs route processing and central configuration and control for the Catalyst 5000 series switch. Routing services can also be provided by an externally attached router.

MLS consists of the following:

- Catalyst 5000 series multilayer LAN switches
- Catalyst RSM, which provides Cisco IOS-based multiprotocol routing and network services

---

NOTE   Cisco 7500, 7200, 4500, and 4700 series routers also support MLS.

- NetFlow Feature Card (NFFC), which is a modular feature-card upgrade for the Catalyst Supervisor Engine III to provide Layer 3 switching

NOTE	The 10/100BaseTX and 100BaseFX Backbone Fast Ethernet Switching modules have onboard hardware that optimizes MLS performance.

Procedures for configuring MLS and NDE on routers are provided in Chapter 13, "Configuring Multilayer Switching."

Terminology

The following terminology is used:

- Multilayer Switching-Switching Engine (MLS-SE)—An NFFC-equipped Catalyst 5000 series switch.

- Multilayer Switching-Route Processor (MLS-RP)—A Cisco router with MLS enabled.

- Multilayer Switching Protocol (MLSP)—The protocol running between the MLS-SE and MLS-RP to enable MLS.

Key MLS Features

Table 12-1 lists the key MLS features.

Table 12-1 *Summary of Key Features*

Feature	Description
Ease of use	Is autoconfigurable and autonomously sets up its Layer 3 flow cache. Its plug-and-play design eliminates the need for you to learn new IP switching technologies.
Transparency	Requires no end-system changes and no renumbering of subnets. It works with DHCP and requires no new routing protocols.
Standards based	Uses IETF standard routing protocols such as OSPF and RIP for route determination. You can deploy MLS in a multivendor network.
Investment protection	Provides a simple feature-card upgrade on the Catalyst 5000 series switches. You can use MLS with your existing chassis and modules. MLS also allows you to use either an integrated RSM or an external router for route processing and Cisco IOS services.
Fast convergence	Allows you to respond to route failures and routing topology changes by performing hardware-assisted invalidation of flow entries.
Resilience	Provides the benefits of HSRP without additional configuration. This feature enables the switches to transparently switch over to the hot standby backup router when the primary router goes offline, eliminating a single point of failure in the network.

Table 12-1 *Summary of Key Features (Continued)*

Feature	Description
Access lists	Allows you to set up access lists to filter or to prevent traffic between members of different subnets. MLS enforces multiple security levels on every packet of the flow at wire speed. It allows you to configure and enforce access control rules on the RSM. Because MLS parses the packet up to the transport layer, it enables access lists to be validated. By providing multiple security levels, MLS enables you to set up rules and control traffic based on IP addresses as well as transport-layer application port numbers.
Accounting and traffic management	Allows you to see data flows as they are switched for troubleshooting, traffic management, and accounting purposes. MLS uses NDE to export the flow statistics. Data collection of flow statistics is maintained in hardware with no impact on switching performance. The records for expired and purged flows are grouped together and exported to applications such as NetSys for network planning, RMON2 traffic management and monitoring, and accounting applications.
Network design simplification	Enables you to speed up your network while retaining the existing subnet structure. It makes the number of Layer 3 hops irrelevant in campus design, enabling you to cope with increases in any-to-any traffic.
Media speed access to server farms	You do not have to centralize servers in multiple VLANs to get direct connections. By providing security on a per-flow basis, you can control access to the servers and filter traffic based on subnet numbers and transport-layer application ports without compromising Layer 3 switching performance.
Faster interworkgroup connectivity	Addresses the need for higher-performance interworkgroup connectivity by intranet and multimedia applications. By deploying MLS, you gain the benefits of both switching and routing on the same platform.

Introduction to Multilayer Switching

Layer 3 protocols, such as IP and Internetwork Packet Exchange (IPX), are connectionless—they deliver each packet independently of each other. However, actual network traffic consists of many end-to-end conversations, or flows, between users or applications.

A flow is a unidirectional sequence of packets between a particular source and destination that share the same protocol and transport-layer information. Communication from a client to a server and from the server to the client are separate flows. For example, Hypertext Transfer Protocol (HTTP) Web packets from a particular source to a particular destination are a separate flow from File Transfer Protocol (FTP) file transfer packets between the same pair of hosts.

Flows can only be based on Layer 3 addresses. This feature allows IP traffic from multiple users or applications to a particular destination to be carried on a single flow only if the destination IP address is used to identify a flow.

The NFFC maintains a Layer 3 switching table (MLS cache) for the Layer 3-switched flows. The cache also includes entries for traffic statistics that are updated in tandem with the switching of packets. After the MLS cache is created, packets identified as belonging to an existing flow, can be Layer 3-switched based on the cached information. The MLS cache maintains flow information for all active flows. When the Layer 3-switching entry for a flow ages out, the flow statistics can be exported to a flow collector application.

Multilayer Switching Implementation

This section provides a step-by-step description of MLS implementation.

NOTE   The MLS-RPs shown in the figures represent either an RSM or an externally attached Cisco router.

---

1 The MLSP informs the Catalyst 5000 series switch of the MLS-RP MAC addresses used on different VLANs and the MLS-RP's routing and access-list changes. Through this protocol, the MLS-RP multicasts its MAC and VLAN information to all MLS-SEs. When the MLS-SE hears the MLSP *hello* message indicating an MLS initialization, the MLS-SE is programmed with the MLS-RP MAC address and its associated VLAN number (see Figure 12-1).

Figure 12-1 *MLS Implementation: Step 1*

MLS-RP multicasts its
MAC addresses and
VLAN number to all
MLS-SEs...

MLS-RP

... all MLS-SEs
program the NFFC
with the MSLP *hello*
message information.

(MLS-SE)

2 In Figure 12-2, host A and host B are located on different VLANs. Host A initiates a data transfer to host B. When host A sends the first packet to the MLS-RP, the MLS-SE recognizes this packet as a *candidate packet* for Layer 3 switching because the MLS-SE has learned the MLS-RP's destination MAC address and VLAN through MLSP. The MLS-SE learns the Layer 3 flow information (such as the destination address, source address, and protocol port numbers) and forwards the first packet to the MLS-RP. A partial MLS entry for this Layer 3 flow is created in the MLS cache.

The MLS-RP receives the packet, looks at its route table to determine how to forward the packet, and applies services such as access control lists and class of service (COS) policy.

The MLS-RP rewrites the MAC header adding a new destination MAC address (host B's) and its own MAC address as the source.

Figure 12-2 *MLS Implementation: Step 2*

Since the Catalyst switch has learned
the MLS-RP's MAC and VLAN information, the
switch starts the MLS process for the Layer 3
flow contained in this packet, the *candidate packet.*

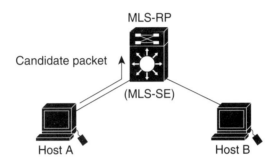

3 The MLS-RP routes the packet to host B. When the packet appears back on the Catalyst 5000 series switch backplane, the MLS-SE recognizes the source MAC address as that of the MLS-RP and that the packet's flow information matches the flow for which it set up a candidate entry. The MLS-SE considers this packet an *enabler packet* and completes the MLS entry (established by the candidate packet) in the MLS cache (see Figure 12-3).

Figure 12-3 *MLS Implementation: Step 3*

The MLS-RP routes this packet to Host B. Since the
MLS-SE has learned both this MLS-RP and the Layer 3
flow in this packet, it completes the MLS entry in the
MLS cache. The first routed packet is called the
enabler packet.

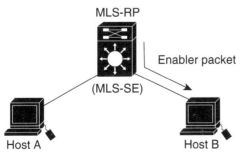

4 After the MLS entry has been completed in Step 3, all Layer 3 packets with the same flow from host A to host B are Layer 3 switched directly inside the switch from host A to host B, bypassing the router (see Figure 12-4). After the Layer 3-switched path is established, the packet from host A is rewritten by the MLS-SE before it is forwarded to host B. The rewritten information includes the MAC addresses, encapsulations (when applicable), and some Layer 3 information.

The resultant packet format and protocol behavior is identical to that of a packet that is routed by the RSM or external Cisco router.

NOTE   MLS is unidirectional. For host B to talk to host A, another Layer 3-switched path needs to be created from host B to host A.

Figure 12-4 *MLS Implementation: Step 4*

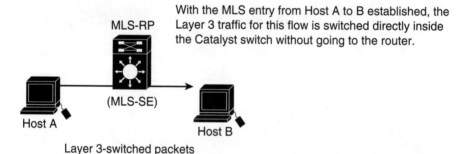

Standard and Extended Access Lists
=========

NOTE   Router interfaces with input access lists *cannot* participate in MLS. However, any input access list can be translated to an output access list to provide the same effect on the interface. For complete details on how input and output access lists affect MLS, see Chapter 13, "Configuring Multilayer Switching."

MLS allows you to enforce access lists on every packet of the flow without compromising MLS performance. When you enable MLS, standard and extended access lists are handled at wire speed by the MLS-SE. Access lists configured on the MLS-RP take effect automatically on the MLS-SE.

Additionally, route topology changes and the addition of access lists are reflected in the switching path of MLS.

Consider the case where an access list is configured on the MLS-RP to deny access from station A to station B. When station A wants to talk to station B, it sends the first packet to the MLS-RP. The MLS-RP receives this packet and checks to see if this packet flow is permitted. If an access control list is configured for this flow, the packet is discarded. Because the first packet for this flow does not return from the MLS-RP, an MLS cache entry is not established by the MLS-SE.

In another case, access lists are introduced on the MLS-RP while the flow is already being Layer 3 switched within the MLS-SE. The MLS-SE immediately enforces security for the affected flow by purging it.

Similarly, when the MLS-RP detects a routing topology change, the appropriate MLS cache entries are deleted in the MLS-SE. The techniques for handling route and access list changes apply to both the RSM and directly attached external routers.

Restrictions on Using IP Router Commands with MLS Enabled

When you issue some Cisco IOS commands, you will affect Multilayer Switching on your router. The commands that will affect MLS are as follows:

- **clear ip-route**—Clears all MLS cache entries for all Catalyst 5000 series switches performing Layer 3 switching for this MLS-RP.

- **ip routing**—The **no** form purges all MLS cache entries and disables MLS on this MLS-RP.

- **ip security** (all forms of this command)—Disables MLS on the interface.

- **ip tcp compression-connections**—Disables MLS on the interface.

- **ip tcp header-compression**—Disables MLS on the interface.

General Guidelines

- When you enable MLS, the RSM or externally attached router continues to handle all non-IP protocols while offloading the switching of IP packets to the MLS-SE.

- Do not confuse MLS with the NetFlow switching supported by Cisco routers. MLS uses both the RSM or directly attached external router and the MLS-SE. With MLS, you *are not* required to use NetFlow switching on the RSM or directly attached external router; any switching path on the RSM or directly attached external router will work (process, fast, and so on).

NOTE	The 10/100BaseTX and 100BaseFX Backbone Fast Ethernet Switching modules for the Catalyst 5000 have onboard hardware that optimizes MLS performance.

Software and Hardware Requirements

MLS requires these software and hardware versions:

- Catalyst 5000 series supervisor engine software

  — Release 4.1(1) or later

- Cisco IOS router software

  — 11.3(2)WA4(4) or later

- Supervisor Engine III with NFFC

- RSM or Cisco 7500, 7200, 4500, or 4700 series router

Guidelines for External Routers

When using an external router, follow these guidelines:

- Cisco recommends one directly attached external router per Catalyst 5000 series switch to ensure that the MLS-SE caches the appropriate flow information from both sides of the routed flow.

- You can use Cisco high-end routers (Cisco 7500, 7200, 4500, and 4700 series) for MLS when they are externally attached to the Catalyst 5000 series switch. You can make the attachment with multiple Ethernets (one per subnet) by using Fast Ethernet with the ISL or with Fast EtherChannel.

- You can connect end hosts through any media (Ethernet, Fast Ethernet, ATM, and FDDI), but the connection between the external router and the Catalyst 5000 series switch must be through standard 10/100 Ethernet interfaces, ISL links, or Fast EtherChannel.

Features That Affect MLS

This section describes how certain features affect MLS.

Access Lists

The following sections describe how access lists affect MLS.

Input Access Lists

Router interfaces with input access lists *cannot* participate in MLS. If you configure an input access list on an interface, all packets for a flow that are destined for that interface go through the router (even if the flow is allowed by the router, it is not Layer 3 switched). Existing flows for that interface get purged and no new flows are cached.

NOTE	Any input access list can be translated to an output access list to provide the same effect on the interface.

Output Access Lists

If an output access list is applied to an interface, the MLS cache entries for that interface are purged. Entries associated with other interfaces are not affected; they follow their normal aging or purging procedures.

Applying an output access list to an interface, when the access list is configured using the **log**, **precedence**, **tos**, or **establish** options, prevents the interface from participating in MLS.

Access List Impact on Flow Masks

Access lists affect the flow mask advertised by an MLS-RP. When there is no access list on any MLS-RP interface, the flow mask mode is destination-ip (the least specific). When there is a standard access list on any of the MLS-RP interfaces, the mode is **source-destination-ip**. When there is an extended access list on any of the MLS-RP interfaces, the mode is **ip-flow** (the most specific).

Reflexive Access Lists

Router interfaces with reflexive access lists *cannot* participate in Layer 3 switching.

IP Accounting

Enabling IP accounting on an MLS-enabled interface disables the IP accounting functions on that interface.

NOTE	To collect statistics for the Layer 3-switched traffic, enable NetFlow Data Export (NDE).

Data Encryption

MLS is disabled on an interface when the data encryption feature is configured on the interface.

Policy Route-Map

MLS is disabled on an interface when a policy route-map is configured on the interface.

TCP Intercept

With MLS interfaces enabled, the TCP intercept feature (enabled in global configuration mode) might not work properly. When you enable the TCP intercept feature, the following message displays:

```
Command accepted, interfaces with mls might cause inconsistent behavior.
```

Network Address Translation

MLS is disabled on an interface when Network Address Translation (NAT) is configured on the interface.

Committed Access Rate

MLS is disabled on an interface when Committed Access Rate (CAR) is configured on the interface.

Maximum Transmission Unit

The MTU for an MLS interface must be the default Ethernet MTU, 1500 bytes.

To change the MTU on an MLS-enabled interface, you must first disable MLS on the interface (enter **no mls rp ip** on the interface). If you attempt to change the MTU with MLS enabled, the following message displays:

```
Need to turn off the mls router for this interface first.
```

If you attempt to enable MLS on an interface that has an MTU value other than the default value, the following message will be displayed:

```
mls only supports interfaces with default mtu size
```

CHAPTER 13

Configuring Multilayer Switching

This chapter describes how to configure your network to perform MLS. For a complete description of the MLS commands see Chapter 14, "Multilayer Switching Commands." For documentation of other commands that appear in this chapter, you can search online at www.cisco.com.

NOTE   The commands and configurations described in this chapter apply only to the devices that provide routing services. Commands and configurations for Catalyst 5000 series switches are not documented here.

Configuring and Monitoring MLS

Perform the tasks in this section to configure your Cisco router for MLS. To ensure a successful MLS configuration, you must also configure the Catalyst switches in your network. Only configuration tasks and commands for routers are described in this chapter.

The following task is required:

● Configuring MLS on a Router

The following tasks are optional:

● Monitoring MLS

● Monitoring MLS for an Interface

● Monitoring MLS Interfaces for VTP Domains

Configuring MLS on a Router

Perform the steps in this section to configure your router(s) for MLS. Depending on your configuration, you might not have to perform all the steps in the procedure. Use the following commands to configure MLS on your router:

Step	Command	Purpose
1	**mls rp ip**	Globally enables MLSP. MLSP is the protocol that runs between the MLS-SE and the MLS-RP.

Step	Command	Purpose
	Use steps 2 through 5 for each interface that will support MLS.	
2	**interface** *type number*	Selects a router interface.
3	**mls rp vtp-domain** [*domain-name*]	Selects the router interface to be Layer 3 switched and then adds that interface to the same VLAN Trunking Protocol (VTP) domain as the switch. This interface is referred to as the MLS interface. This command is required only if the Catalyst switch is in a VTP domain.
4	**mls rp vlan-id** [*vlan-id-num*]	Assigns a VLAN ID to the MLS interface. MLS requires that each interface has a VLAN ID. This step is not required for RSM VLAN interfaces or ISL-encapsulated interfaces.
5	**mls rp ip**	Enables each MLS interface.
6	**mls rp management-interface**	Selects one MLS interface as a management interface. MLSP packets are sent and received through this interface. This can be any MLS interface connected to the switch.

NOTE   The interface-specific commands in this section apply only to Ethernet, Fast Ethernet, VLAN, and Fast EtherChannel interfaces on the Catalyst RSM/VIP2 or directly attached external router.

Use the following command to globally disable MLS on the router:

Command	Purpose
no mls rp ip	Disables MLS on the router.

Monitoring MLS

Use the **show mls rp** command to display MLS details, including specifics for MLSP. Displays include

- MLS status (enabled or disabled) for switch interfaces and subinterfaces
- Flow mask used by this MLS-enabled switch when creating Layer 3 switching entries for the router
- Current settings of the keepalive timer, retry timer, and retry count
- MLSP-ID used in MLSP messages
- List of interfaces in all VTP domains that are enabled for MLS

Command	Purpose
show mls rp	Shows MLS details for all interfaces.

After entering this command, you see this display:

```
router# show mls rp
multilayer switching is globally enabled
mls id is 00e0.fefc.6000
mls ip address 10.20.26.64
mls flow mask is ip-flow

vlan domain name: WBU
   current flow mask: ip-flow
   current sequence number: 80709115
   current/maximum retry count: 0/10
   current domain state: no-change
   current/next global purge: false/false
   current/next purge count: 0/0
   domain uptime: 13:03:19
   keepalive timer expires in 9 seconds
   retry timer not running
   change timer not running
   fcp subblock count = 7

   1 management interface(s) currently defined:
      vlan 1 on Vlan1

   7 mac-vlan(s) configured for multi-layer switching:

      mac 00e0.fefc.6000
         vlan id(s)
         1    10   91   92   93   95   100

   router currently aware of following 1 switch(es):
      switch id 0010.1192.b5ff

router#
```

Monitoring MLS for an Interface

Use the following command to show MLS information for a specific interface:

Command	Purpose
show mls rp [*interface*]	Shows MLS details for a specific interface.

After entering this command, you see this display:

```
router# show mls rp int vlan 10
mls active on Vlan10, domain WBU
router#
```

Monitoring MLS Interfaces for VTP Domains

Command	Purpose
show mls rp vtp-domain [*domain-name*]	Shows MLS interfaces for a specific VTP domain.

After entering this command, you see this display:

```
router# show mls rp vtp-domain WBU
vlan domain name: WBU
   current flow mask: ip-flow
   current sequence number: 80709115
   current/maximum retry count: 0/10
   current domain state: no-change
   current/next global purge: false/false
   current/next purge count: 0/0
   domain uptime: 13:07:36
   keepalive timer expires in 8 seconds
   retry timer not running
   change timer not running
   fcp subblock count = 7

   1 management interface(s) currently defined:
     vlan 1 on Vlan1

   7 mac-vlan(s) configured for multi-layer switching:

     mac 00e0.fefc.6000
       vlan id(s)
       1    10   91   92   93   95   100

   router currently aware of following 1 switch(es):
     switch id 0010.1192.b5ff

router#
```

Configuring NetFlow Data Export

NOTE   You need to enable NDE only if you are going to export MLS cache entries to a data collection application.

Perform the task in this section to configure your Cisco router for NDE. To ensure a successful NDE configuration, you must also configure the Catalyst switch. Only configuration tasks and commands for routers are described in this chapter.

Perform the following tasks to configure NDE on your network. The first task is for the router, and the remaining tasks are for the switch:

- Specifying a NetFlow Data Export Address on the Router
- Specifying a NetFlow Data Export Collector
- Enabling NetFlow Data Export on the Switch
- Specifying a Filter for NDE Flow from the Switch

Specifying a NetFlow Data Export Address on the Router

Use the following command to specify an NDE address on the router:

Command	Purpose
mls rp nde-address *ip-address*	Specifies an NDE IP address for the router doing the Layer 3 switching. The router and the Catalyst 5000 series switch use the NDE IP address when sending MLS statistics to a data collection application.

Multilayer Switching Configuration Examples

In these examples, VLAN interfaces 1 and 3 are in VTP domain Engineering. The management interface is configured on the VLAN 1 interface. Only information relevant to MLS is shown in the following configurations:

- Router Configuration without Access Lists
- Router Configuration with Standard Access List
- Router Configuration with Extended Access List

Router Configuration without Access Lists

This sample configuration shows a router configured without access lists on any of the VLAN interfaces. The flow mask is configured to be **destination-ip**:

```
router# more system:running-config
Building configuration...

Current configuration:
.
.
.
mls rp ip

interface Vlan1
 ip address 172.20.26.56 255.255.255.0
```

```
 mls rp vtp-domain Engineering
 mls rp management-interface
 mls rp ip

interface Vlan2
 ip address 128.6.2.73 255.255.255.0

interface Vlan3
 ip address 128.6.3.73 255.255.255.0
 mls rp vtp-domain Engineering
 mls rp ip
 .
 .
 end
router#
router# show mls rp
multilayer switching is globally enabled
mls id is 0006.7c71.8600
mls ip address 172.20.26.56
mls flow mask is destination-ip

number of domains configured for mls 1
vlan domain name: Engineering
   current flow mask: destination-ip
   current sequence number: 82078006
   current/maximum retry count: 0/10
   current domain state: no-change
   current/next global purge: false/false
   current/next purge count: 0/0
   domain uptime: 02:54:21
   keepalive timer expires in 11 seconds
   retry timer not running
   change timer not running

   1 management interface(s) currently defined:
      vlan 1 on Vlan1

   2 mac-vlan(s) configured for multi-layer switching:

      mac 0006.7c71.8600
         vlan id(s)
         1     3

   router currently aware of following 1 switch(es):
      switch id 00e0.fe4a.aeff
router#
```

Router Configuration with Standard Access List

This configuration is the same as the previous example but with a standard access list configured on the VLAN 3 interface. The flow mask changes to **source-destination-ip**:

```
.
interface Vlan3
 ip address 128.6.3.73 255.255.255.0
 ip access-group 2 out
 mls rp vtp-domain Engineering
 mls rp ip
.

router# show mls rp
multilayer switching is globally enabled
mls id is 0006.7c71.8600
mls ip address 172.20.26.56
mls flow mask is source-destination-ip

number of domains configured for mls 1
vlan domain name: Engineering
   current flow mask: source-destination-ip
   current sequence number: 82078007
   current/maximum retry count: 0/10
   current domain state: no-change
   current/next global purge: false/false
   current/next purge count: 0/0
   domain uptime: 02:57:31
   keepalive timer expires in 4 seconds
   retry timer not running
   change timer not running

   1 management interface(s) currently defined:
      vlan 1 on Vlan1

   2 mac-vlan(s) configured for multi-layer switching:

      mac 0006.7c71.8600
         vlan id(s)
         1    3

   router currently aware of following 1 switch(es):
      switch id 00e0.fe4a.aeff

router#
```

Router Configuration with Extended Access List

This configuration is the same as the previous examples but with an extended access list configured on the VLAN 3 interface. The flow mask changes to **ip-flow**:

```
.
interface Vlan3
 ip address 128.6.3.73 255.255.255.0
 ip access-group 101 out
 mls rp vtp-domain Engineering
 mls rp ip
.


router# show mls rp
multilayer switching is globally enabled
mls id is 0006.7c71.8600
mls ip address 172.20.26.56
mls flow mask is ip-flow

number of domains configured for mls 1
vlan domain name: Engineering
   current flow mask: ip-flow
   current sequence number: 82078009
   current/maximum retry count: 0/10
   current domain state: no-change
   current/next global purge: false/false
   current/next purge count: 0/0
   domain uptime: 03:01:52
   keepalive timer expires in 3 seconds
   retry timer not running
   change timer not running

   1 management interface(s) currently defined:
      vlan 1 on Vlan1

   2 mac-vlan(s) configured for multi-layer switching:

      mac 0006.7c71.8600
        vlan id(s)
        1    3

   router currently aware of following 1 switch(es):
      switch id 00e0.fe4a.aeff

router#
```

CHAPTER 14

Multilayer Switching Commands

This chapter documents commands used to configure MLS in Cisco IOS software. For guidelines on configuring MLS, refer to Chapter 13, "Configuring Multilayer Switching."

NOTE   Beginning with Cisco IOS Release 11.3, all commands supported on the Cisco 7500 series routers are also supported on Cisco 7000 series routers.

---

mls rp ip

To enable Multilayer Switching Protocol (MLSP), use the **mls rp ip** global configuration command. MLSP is the protocol that runs between the switches and routers. Use the **no** form of this command to disable MLS.

> **mls rp ip**
> **no mls rp ip**

Syntax Description

There are no arguments or keywords for this command.

Default

The default is MLS disabled.

Command Mode

Global configuration

Usage Guidelines

This command first appeared in Cisco IOS Release 11.3(3) WA4(4).

Use this command to enable MLS, either globally or on a specific interface.

Example

The following example enables MLS:

```
mls rp ip
```

Related Commands

You can search online at www.cisco.com to find documentation of related commands.

mls rp management-interface
mls rp nde-address
mls rp vlan-id
mls rp vtp-domain
show mls rp
show mls rp vtp-domain

mls rp management-interface

To designate an interface as the management interface for MLSP packets, use the **mls rp management-interface** interface configuration command. Use the **no** version of the command to remove an interface as the management interface.

>   **mls rp management-interface**
>   **no mls rp management-interface**

Syntax Description

This command has no arguments or keywords.

Default

There is no default management interface.

Command Mode

Interface configuration

Usage Guidelines

This command first appeared in Cisco IOS Release 11.3(3) WA4(4).

Use this command to designate an interface as the MLSP management interface. You must specify a router interface as a management interface. If you do not specify an interface, MLSP packets *will not* be sent or received.

The management interface can be any MLS interface connected to the Catalyst 5000 series switch. Specifying more than one interface is not necessary.

Example

The following example sets the current interface as the management interface:

```
mls rp management-interface
```

Related Commands

You can search online at www.cisco.com to find documentation of related commands.

mls rp ip
mls rp nde-address
mls rp vlan-id
mls rp vtp-domain
show mls rp
show mls rp vtp-domain

mls rp nde-address

To specify an NDE address, use the **mls rp nde-address** global configuration command.

> **mls rp nde-address** *ip-address*

Syntax	Description
ip-address	NDE IP address.

Command Mode

Global configuration

Usage Guidelines

This command first appeared in Cisco IOS Release 11.3(3) WA4(4).

Use this command on an RP to specify the NetFlow Data Export address for a router. If you *do not* specify an NDE IP address for the MLS-RP, the MLS-RP automatically selects one of its interface's IP addresses and uses that IP address as its NDE IP address *and* its **mls ip address**.

Example

The following example sets the NDE address to 170.25.2.1:

```
mls rp nde-address 170.25.2.1
```

Related Commands

You can search online at www.cisco.com to find documentation of related commands.

mls rp ip
mls rp management-interface
mls rp vlan-id
mls rp vtp-domain
show mls rp
show mls rp vtp-domain

mls rp vlan-id

To assign a VLAN ID, use the **mls ip vlan-id** interface configuration command.

> **mls rp vlan-id** *vlan-id-num*

Syntax	Description
vlan-id-num	VLAN identification number.

Command Mode

Interface configuration

Usage Guidelines

This command first appeared in Cisco IOS Release 11.3(3) WA4(4).

Use this command to assign a VLAN ID to an interface. RSM VLAN interfaces or ISL-encapsulated interfaces do not require the VLAN ID to be assigned.

Example

The following example assigns a VLAN ID of 23 to the current interface:

```
mls rp vlan-id 23
```

Related Commands

You can search online at www.cisco.com to find documentation of related commands.

mls rp ip
mls rp management-interface
mls rp nde-address
mls rp vtp-domain
show mls rp
show mls rp vtp-domain

mls rp vtp-domain

To select the router interface to be Layer 3 switched and then add that interface to a VTP domain, use the **mls rp vtp-domain** interface configuration command.

mls rp vtp-domain *domain-name*

Syntax

domain-name

Description

VTP domain name

Command Mode

EXEC

Usage Guidelines

This command first appeared in Cisco IOS Release 11.3(3) WA4(4).

This command is required only if the Catalyst switch is in a VTP domain. For an ISL interface, you can enter this command only on the primary interface. All subinterfaces that are part of the primary interface inherit the primary's VTP domain.

Example

The following example adds the interface to the engineering VTP domain:

```
mls rp vtp-domain engineering
```

Related Commands

You can search online at www.cisco.com to find documentation of related commands.

mls rp ip
mls rp management-interface
mls rp nde-address
mls rp vlan-id
show mls rp
show mls rp vtp-domain

show mls rp

To display MLS details, including specifics for MLSP, use the **show mls rp** EXEC command.

       show mls rp [*interface*]

Syntax	Description
interface	(Optional) Displays information for one interface. Without this argument, detailed view of all interfaces are displayed.

Command Mode

EXEC

Usage Guidelines

This command first appeared in Cisco IOS Release 11.3(3) WA4(4).

Sample Display

The following is sample output for the **show mls rp** command:

```
router# show mls rp
multilayer switching is globally enabled
mls id is 00e0.fefc.6000
mls ip address 10.20.26.64
mls flow mask is ip-flow
vlan domain name: WBU
```

```
current flow mask: ip-flow
current sequence number: 80709115
current/maximum retry count: 0/10
current domain state: no-change
current/next global purge: false/false
current/next purge count: 0/0
domain uptime: 13:03:19
keepalive timer expires in 9 seconds
retry timer not running
change timer not running
fcp subblock count = 7

1 management interface(s) currently defined:
   vlan 1 on Vlan1

7 mac-vlan(s) configured for multi-layer switching:

   mac 00e0.fefc.6000
     vlan id(s)
     1    10   91   92   93   95   100

router currently aware of following 1 switch(es):
   switch id 0010.1192.b5ff
```

The following is sample output for the **show mls rp** command for a specific interface:

```
router# show mls rp int vlan 10
mls active on Vlan10, domain WBU
router#
```

Related Commands

You can search online at www.cisco.com to find documentation of related commands.

mls rp ip
mls rp management-interface
mls rp nde-address
mls rp vlan-id
mls rp vtp-domain
show mls rp vtp-domain

show mls rp vtp-domain

To show MLS interfaces for a specific VTP domain, use the **show mls rp vtp-domain** EXEC command.

show mls rp vtp-domain [*domain-name*]

Syntax	Description
domain-name	VTP domain name.

Command Mode

EXEC

Usage Guidelines

This command first appeared in Cisco IOS Release 11.3(3) WA4(4).

Sample Display

The following example is sample output from the **show mls rp vtp-domain** command:

```
router# show mls rp vtp-domain WBU
vlan domain name: WBU
   current flow mask: ip-flow
   current sequence number: 80709115
   current/maximum retry count: 0/10
   current domain state: no-change
   current/next global purge: false/false
   current/next purge count: 0/0
   domain uptime: 13:07:36
   keepalive timer expires in 8 seconds
   retry timer not running
   change timer not running
   fcp subblock count = 7

   1 management interface(s) currently defined:
      vlan 1 on Vlan1

   7 mac-vlan(s) configured for multi-layer switching:

      mac 00e0.fefc.6000
         vlan id(s)
         1    10    91    92    93    95    100

   router currently aware of following 1 switch(es):
      switch id 0010.1192.b5ff
```

Related Commands

You can search online at www.cisco.com to find documentation of related commands.

mls rp ip
mls rp management-interface
mls rp nde-address
mls rp vlan-id
mls rp vtp-domain
show mls rp

PART **VI**

Multicast Distributed Switching

CHAPTER 15

Configuring Multicast Distributed Switching

This chapter describes the required and optional tasks for configuring multicast distributed switching (MDS). For a complete description of MDS commands used in this chapter, refer to Chapter 16, "Multicast Distributed Switching Commands." For documentation of other commands that appear in this chapter, you can search online at www.cisco.com.

Prior to MDS, IP multicast traffic was always switched at the RP in the RSP-based platforms. Starting with Cisco IOS Release 11.2 GS, IP multicast traffic can be distributed switched on RSP-based platforms with VIPs. Furthermore, MDS is the only multicast switching method on the Cisco 12000 Gigabit Switched Router (GSR), starting with Cisco IOS Release 11.2(11)GS.

Switching multicast traffic at the RP had disadvantages:

- The load on the RP increased. This affected important route updates and calculations (for BGP, among others) and could stall the router if the multicast load was significant.

- The net multicast performance was limited to what a single RP could switch.

MDS solves these problems by performing distributed switching of multicast packets received at the line cards (VIPs in the case of RSP, and line cards in the case of GSR). The line card is the interface card that houses the VIPs (in the case of RSP) and the GSR line card (in the case of GSR). MDS is accomplished by using a forwarding data structure called a Multicast Forwarding Information Base (MFIB), which is a subset of the routing table. A copy of MFIB runs on each line card and is always kept up-to-date with the RP's MFIB table. In the case of RSP, packets received on non-VIP IPs are switched by the RP.

MDS can work in conjunction with CEF, unicast distributed fast switching (DFS), or flow switching.

Configure MDS

This section describes the tasks to configure MDS. The first task is required.

- Enabling MDS

- Monitoring and Maintaining MDS

Enabling MDS

To enable MDS, you must enable it globally and on at least one interface because MDS is an attribute of the interface. Use the following commands, beginning in global configuration mode:

Step	Command	Purpose
1	**ip multicast-routing distributed**	Enables MDS globally.
2	**interface** *type number*	Configures an interface.
3	**ip route-cache distributed**	Enables distributed switching on the RSP. (This step is required on the RSP platform only.)
4	**ip mroute-cache distributed**	Enables MDS on the interface.
5		Repeat Steps 2 through 4 for each interface that you want to perform MDS.

NOTE	When you enable an interface to perform distributed switching of incoming multicast packets, you are configuring the physical interface, not the logical interface (subinterface). All subinterfaces are included in the physical interface.

Monitoring and Maintaining MDS

To maintain MDS on the line cards, use the following command in EXEC mode:

Command	Purpose
clear ip mds forwarding	Clears the line card's MFIB table and resynchronizes with the RP.

To maintain MDS on the RP, use the following commands in EXEC mode:

Command	Purpose
clear ip mroute {* | *group* [*source*]}	Clears multicast routes and counts.
clear ip pim interface count	Clears all packet counts on the line cards.

To monitor MDS on the line cards, use the following commands in EXEC mode (remember that to reach a line card's console, enter **attach** `slot#`, using the slot number where the line card resides):

Command	Purpose
show ip mds forwarding [*group-address*] [*source-address*]	Displays the MFIB table, forwarding information, related flags, and counts.
show ip mds summary	Displays a summary of the MFIB.

To monitor MDS on the RP, use the following commands in EXEC mode:

Command	Purpose
show ip mds stats [switching | linecard]	Displays switching statistics or line card statistics for MDS.
show ip mds interface	Displays the status of MDS interfaces.
show ip pim interface [*type number*] **count**	Displays switching counts for unicast DFS and other fast switching statistics.
show ip mcache [*group* [*source*]]	Displays the contents of the IP fast-switching cache.
show interface stats	Displays numbers of packets that were process switched, fast switched, and distributed switched.

Configuration Example

The following example enables MDS. The command **ip route-cache distributed** is needed on the RSP only, not on the GSR.

```
ip multicast-routing distributed
 interface pos 1/0/0
 ip route-cache distributed
 ip mroute-cache distributed
```

CHAPTER 16

Multicast Distributed Switching Commands

This chapter documents commands used to configure MDS in Cisco IOS software. For guidelines on configuring MDS, refer to Chapter 15, "Configuring Multicast Distributed Switching."

NOTE   Beginning with Cisco IOS Release 11.3, all commands supported on the Cisco 7500 series routers are also supported on Cisco 7000 series routers.

---

clear ip mds forwarding

To clear all routes from a line card's MFIB table and resynchronize it with the RP, use the **clear ip mds forwarding** EXEC command.

> **clear ip mds forwarding**

Syntax Description

This command has no arguments or keywords.

Command Mode

EXEC

Usage Guidelines

This command first appeared in Cisco IOS Release 11.2(11)GS.

Use this command on a line card of a Cisco 7500 or Cisco 12000.

Example

The following example clears the line card's MFIB table:

```
clear ip mds forwarding
```

Related Commands

You can search online at www.cisco.com to find documentation of related commands.

clear ip pim interface count

clear ip pim interface count

To clear all line card counts or packet counts, use the **clear ip pim interface count** EXEC command.

> **clear ip pim interface count**

Syntax Description

This command has no arguments or keywords.

Command Mode

EXEC

Usage Guidelines

This command first appeared in Cisco IOS Release 11.2(11)GS.

Use this command on an RP to delete all MDS statistics for the entire router.

Example

The following example clears all the line card packet counts:

```
clear ip pim interface count
```

Related Commands

You can search online at www.cisco.com to find documentation of related commands.

clear ip mds forwarding

ip mroute-cache

To configure IP multicast fast switching or MDS, use the **ip mroute-cache** interface configuration command. To disable either of these features, use the **no** form of this command.

>**ip mroute-cache [distributed]**
>**no ip mroute-cache [distributed]**

Syntax	Description
distributed	(Optional) Enables MDS on the interface. In the case of RSP, this keyword is optional; if it is omitted, fast switching occurs. On the GSR, this keyword is required because the GSR does only distributed switching.

Default

On the RSP, IP multicast fast switching is enabled; MDS is disabled.
On the GSR, MDS is disabled.

Command Mode

Interface configuration

Usage Guidelines

This command first appeared in Cisco IOS Release 11.0. The **distributed** keyword first appeared in Release 11.2(11)GS.

On the RSP

If multicast fast switching is disabled on an incoming interface for a multicast routing table entry, the packet will be sent at process level for all interfaces in the outgoing interface list.

If multicast fast switching is disabled on an outgoing interface for a multicast routing table entry, the packet is process-level switched for that interface, but may be fast switched for other interfaces in the outgoing interface list.

When multicast fast switching is enabled (like unicast routing), debug messages are not logged. If you want to log debug messages, disable fast switching.

If MDS is not enabled on an incoming interface that is capable of MDS, incoming multicast packets will not be distributed switched; they will be fast switched at the RP as before. Also, if the incoming interface is not capable of MDS, packets will be fast switched or process switched at the RP as before.

If MDS is enabled on the incoming interface, but at least one of the outgoing interfaces cannot fast switch, packets will be process switched. So it is a good idea not to disable fast switching on any interface when MDS is enabled.

On the GSR

On the GSR, all interfaces should be configured for MDS because that is the only switching mode.

Examples

The following example enables IP multicast fast switching on the interface:

```
ip mroute-cache
```

The following example disables IP multicast fast switching on the interface:

```
no ip mroute-cache
```

The following example enables MDS on the interface:

```
ip mroute-cache distributed
```

The following example disables MDS and IP multicast fast switching on the interface:

```
no ip mroute-cache distributed
```

ip multicast-routing

To enable IP multicast routing or MDS, use the **ip multicast-routing** global configuration command. To disable IP multicast routing and MDS, use the **no** form of this command.

> **ip multicast-routing** [**distributed**]
> **no ip multicast-routing**

Syntax	Description
distributed	(Optional) Enables MDS.

Default

Disabled

Command Mode

Global configuration

Usage Guidelines

This command first appeared in Cisco IOS Release 10.0. The **distributed** keyword first appeared in Release 11.2(11)GS.

When IP multicast routing is disabled, the Cisco IOS software does not forward any multicast packets.

Examples

The following example enables IP multicast routing:

```
ip multicast-routing
```

The following example disables IP multicast routing and MDS:

```
no ip multicast-routing
```

Related Commands

You can search online at www.cisco.com to find documentation of related commands.

ip pim

show interface stats

To display numbers of packets that were process switched, fast switched, and distributed switched, use the **show interface stats** EXEC command.

> **show interface** *type number* **stats**

Syntax   Description

type number   Interface type and number about which to display statistics.

Command Mode

EXEC

Usage Guidelines

This command first appeared in Cisco IOS Release 11.0.

Use this command on the RP.

Sample Display

The following is sample output from the **show interface stats** command:

```
Router# show interface fddi 3/0/0 stats

Fddi3/0/0
        Switching path   Pkts In   Chars In   Pkts Out   Chars Out
            Processor    3459994  1770812197    4141096  1982257456
           Route cache  10372326  3693920448     439872   103743545
      Distributed cache 19257912  1286172104   86887377  1184358085
                 Total  33090232  2455937453   91468345  3270359086
```

Table 16-1 describes the fields in the display of the **show interface stats** command.

Table 16-1 *show interface stats Field Descriptions*

Field	Description
Fddi3/0/0	Interface for which information is shown.
Switching path	Column heading for the various switching paths below it.
Pkts In	Number of packets received in each switching mechanism.
Chars In	Number of characters received in each switching mechanism.
Pkts Out	Number of packets sent out each switching mechanism.
Chars Out	Number of characters sent out each switching mechanism.

show ip mcache

To display the contents of the IP multicast fast-switching cache, use the **show ip mcache** EXEC command.

>   **show ip mcache** [*group* [*source*]]

Syntax	Description
group	(Optional) Displays the fast-switching cache for the single group. The *group* argument can be either a Class D IP address or a DNS name.
source	(Optional) If *source* is also specified, displays a single multicast cache entry. The *source* argument can be either a unicast IP address or a DNS name.

Command Mode

EXEC

Usage Guidelines

This command first appeared in Cisco IOS Release 11.0.

Use this command on the RP.

Sample Displays

The following is sample output from the **show ip mcache** command. This entry shows a specific source (wrn-source 204.62.246.73) sending to the World Radio Network group (224.2.143.24).

```
Router> show ip mcache wrn wrn-source

IP Multicast Fast-Switching Cache
(204.62.246.73/32, 224.2.143.24), Fddi0, Last used: 00:00:00
  Ethernet0      MAC Header: 01005E028F1800000C1883D30800
  Ethernet1      MAC Header: 01005E028F1800000C1883D60800
  Ethernet2      MAC Header: 01005E028F1800000C1883D40800
  Ethernet3      MAC Header: 01005E028F1800000C1883D70800
```

Table 16-2 describes the significant fields in the display of the **show ip mcache** command.

Table 16-2 *show ip mcache Field Description*

Field	Description
204.62.246.73	Source address.
224.2.143.24	Destination address.
Fddi0	Incoming or expected interface on which the packet should be received.
Last used:	Latest time the entry was accessed for a packet that was successfully fast switched: • *semi-fast* indicates that the first part of the outgoing interface list is fast switched and the rest of the list is process-level switched. • *mds* indicates that MDS is being used instead of the fast cache. • *never* indicates that the fast cache entry is not used (it is process switched).
Ethernet0 MAC Header:	Outgoing interface list and respective MAC header that is used when rewriting the packet for output. If the interface is a tunnel, the MAC header will show the real next hop MAC header and then, in parentheses, the real interface name.

The following is sample output from the **show ip mcache** command when MDS is in effect:

```
Router# show ip mcache

IP Multicast Fast-Switching Cache
(*, 224.2.170.73), Fddi3/0/0, Last used: mds
  Tunnel3      MAC Header: 5000602F9C150000603E473F60AAAA030000000800 (Fddi3/0/0)
  Tunnel0      MAC Header: 5000602F9C150000603E473F60AAAA030000000800 (Fddi3/0/0)
  Tunnel1      MAC Header: 5000602F9C150000603E473F60AAAA030000000800 (Fddi3/0/0)
```

show ip mds forwarding

On a line card, to display the MFIB table and forwarding information for MDS, use the **show ip mds forwarding** EXEC command.

> **show ip mds forwarding** [*group-address*] [*source-address*]

Syntax	Description
group-address	(Optional) Address of the IP multicast group for which to display the MFIB table.
source-address	(Optional) Address of the source of IP multicast packets for which to display the MFIB table.

Command Mode

EXEC

Usage Guidelines

This command first appeared in Cisco IOS Release 11.2(11)GS.

Use this command on the line card. This command displays the MFIB table, forwarding information, and related flags and counts.

NOTE   To reach a line card's console, enter **attach** *slot#* (the slot number where the line card resides).

On a GSR only, line card commands can be executed from the RP by using the following syntax:
execute [**slot** *slot-number* | **all**] *command*

command is any of the line card **show** commands, such as **show ip mds summary** and **show ip mds forward**.

Sample Display

The following is sample output from the **show ip mds forwarding** command:

```
Router# show ip mds forwarding
IP multicast MDFS forwarding information and statistics:
Flags: N - Not MDFS switchable, F - Not all MDFS switchable, O - OIF Null
       R - In-ratelimit, A - In-access, M - MTU mismatch, P - Register set
```

```
Interface state: Interface, Next-Hop, Mac header

(*, 224.2.170.73),
  Incoming interface: Null
  Pkts: 0, last used: never, Kbps: 0, fast-flags: N
  Outgoing interface list: Null

(128.97.62.86, 224.2.170.73) [31]
  Incoming interface: Fddi3/0/0
  Pkts: 3034, last used: 00:00:00, Kbps: 0, fast-flags: M
  Outgoing interface list:
```

Table 16-3 describes the significant fields in the display of the **show ip mds forwarding** command.

Table 16-3 *show ip mds forwarding Field Descriptions*

Field	Description
(128.97.62.86, 224.2.170.73) [31])	Source and group addresses. Number in [ ] is the hash bucket for the route.
Incoming interface	Expected interface for a multicast packet from the source. If the packet is not received on this interface, it is discarded.
Pkts	Total number of packets switched by that entry.
last used	Time when this MFIB entry was used to switch a packet.
Kbps	Kilobits per second of the switched traffic.
Outgoing interface list	Interfaces through which packets will be forwarded.

show ip mds interface

To display the status of MDS interfaces, use the **show ip mds interface** EXEC command.

show ip mds interface

Syntax Description

This command has no arguments or keywords.

Command Mode

EXEC

Usage Guidelines

This command first appeared in Cisco IOS Release 11.2(11)GS.

Use this command on the RP.

Sample Display

The following is sample output from the **show ip mds interface** command:

```
Router# show ip mds interface

Ethernet1/0/0 is up, line protocol is up
Ethernet1/0/1 is up, line protocol is up
Fddi3/0/0 is up, line protocol is up
FastEthernet3/1/0 is up, line protocol is up
```

Table 16-4 describes the fields in the display of the **show ip mds interface** command.

Table 16-4 *show ip mds interface Field Descriptions*

Field	Description
Ethernet1/0/0 is up	Status of interface.
line protocol is up	Status of line protocol.

show ip mds stats

To display switching statistics or line card statistics for MDS, use the **show ip mds stats** EXEC command.

> **show ip mds stats [switching | linecard]**

Syntax	Description
switching	(Optional) Displays switching statistics.
linecard	(Optional) Displays line card statistics.

Command Mode

EXEC

Usage Guidelines

This command first appeared in Cisco IOS Release 11.2(11)GS.

Use this command on the RP.

Sample Displays

The following is sample output from the **show ip mds stats** command with the **switching** keyword:

```
Router# show ip mds stats switching

Slot Total        Switched    Drops      RPF        Punts      Failures
                                                               (switch/clone)
 1   0            0           0          0          4          0/0
 3   20260925     18014717    253        93         2247454    1/0
```

Table 16-5 describes the fields in the display of the **show ip mds stats** command.

Table 16-5 *show ip mds stats switching Field Descriptions*

Field	Description
Slot	Slot number for the line card.
Total	Total number of packets received.
Switched	Total number of packets switched.
Drops	Total number of packets dropped.
RPF	Total number of packets that failed RPF lookup.
Punts	Total number of packets sent to the RP because the line card could not switch them.
Failures (switch/clone)	Times that the RP tried to switch but failed due to lack of resources/clone for RSP only; failed to get a packet clone.

The following is sample output from the **show ip mds stats** command with the **linecard** keyword:

```
Router# show ip mds stats linecard

Slot    Status    IPC(seq/max) Q(high/route)  Reloads
 1      active    10560/10596     0/0             9
 3      active    11055/11091     0/0             9
```

Table 16-6 describes the fields in the display of the **show ip mds stats** command.

Table 16-6 *show ip mds stats linecard Field Descriptions*

Field	Description
Slot	VIP card slot.
IPC (seq/max)	Interprocess communication of packets sent from the RP to the VIP.
Reloads	Number of times the image on the VIP was reloaded.

show ip mds summary

To display a summary of the MFIB table for MDS, use the **show ip mds summary** EXEC command.

show ip mds summary

Syntax Description

This command has no arguments or keywords.

Command Mode

EXEC

Usage Guidelines

This command first appeared in Cisco IOS Release 11.2(11)GS.

Use this command on a line card. On a GSR only, line card commands can be executed from the RP by using the following syntax:

execute [**slot** *slot-number* | **all**] *command*

command is any of the line card **show** commands, such as **show ip mds summary** and **show ip mds forward**.

Sample Display

The following is sample output from the **show ip mds summary** command:

```
Router# show ip mds summary

IP multicast MDFS forwarding information and statistics:
Flags: N - Not MDFS switchable, F - Not all MDFS switchable, O - OIF Null
       R - In-ratelimit, A - In-access, M - MTU mismatch, P - Register set

Interface state: Interface, Next-Hop, Mac header

(*, 224.2.170.73),
  Incoming interface: Null
  Pkts: 0, last used: never, Kbps: 0, fast-flags: N
(128.97.62.86, 224.2.170.73) [31]
  Incoming interface: Fddi3/0/0
  Pkts: 3045, last used: 00:00:03, Kbps: 0, fast-flags: M
(128.223.3.7, 224.2.170.73) [334]
  Incoming interface: Fddi3/0/0
  Pkts: 0, last used: never, Kbps: 0, fast-flags: M
```

Table 16-7 describes the fields in the display of the **show ip mds summary** command.

Table 16-7   *show ip mds summary Field Descriptions*

Field	Description
(128.97.62.86, 224.2.170.73) [31]	Source and group addresses. Number in [ ] is the hash bucket for the route.
Incoming interface	Expected interface for a multicast packet from the source. If the packet is not received on this interface, it is discarded.
Pkts	Total number of packets switched by that entry.
last used	Time when this MFIB entry was used to switch a packet.
Kbps	Kilobits per second of the switched traffic.

show ip pim interface count

To display switching counts for MDS and other fast-switching statistics, use the **show ip pim interface count** EXEC command.

> **show ip pim interface** [*type number*] **count**

Syntax

Description

type number   (Optional) Interface type and number. If these arguments are specified, information is displayed about that interface only.

Command Mode
EXEC

Usage Guidelines
This command first appeared in Cisco IOS Release 11.2(11)GS.

Use this command on the RP.

Sample Display

The following is sample output from the **show ip pim interface count** command:

```
Router# show ip pim interface count

Address           Interface          FS   Mpackets In/Out
128.223.224.8     Ethernet1/0/0      D    4/0
128.223.225.1     Ethernet1/0/1      D    0/0
128.223.222.8     Fddi3/0/0          D    20182993/56931
128.223.156.1     FastEthernet3/1/0  D    59991/462385
137.39.26.98      Tunnel0            *    394681/10686513
128.223.90.13     Tunnel1            *    517821/7185605
128.223.90.25     Tunnel3            *    26282/20027641
128.223.90.29     Tunnel4            *    2415/8688961
```

Table 16-8 describes the fields in the display of the **show ip pim interface count** command.

Table 16-8 *show ip pim interface count Field Descriptions*

Field	Description
Address	Source address of the IP multicast packet.
Interface	Interface on which the packets are arriving.
FS	D indicates that the packets were distributed switched.
Mpackets In/Out	Number of multicast packets received/number of multicast packets sent out.

Virtual LANs

Routing Between Virtual LANs Overview

This chapter provides an overview of virtual LANs (VLANs). It describes the encapsulation protocols used for routing between VLANs and provides some basic information about designing VLANs.

What Is a Virtual LAN?

A VLAN is a switched network that is logically segmented on an organizational basis, by functions, project teams, or applications rather than on a physical or geographical basis. For example, all workstations and servers used by a particular workgroup team can be connected to the same VLAN, regardless of their physical connections to the network or the fact that they might be intermingled with other teams. Reconfiguration of the network can be done through software rather than by physically unplugging and moving devices or wires.

A VLAN can be thought of as a broadcast domain that exists within a defined set of switches. A VLAN consists of a number of end systems, either hosts or network equipment (such as bridges and routers), connected by a single bridging domain. The bridging domain is supported on various pieces of network equipment; for example, LAN switches that operate bridging protocols between them with a separate bridge group for each VLAN.

VLANs are created to provide the segmentation services traditionally provided by routers in LAN configurations. VLANs address scalability, security, and network management. Routers in VLAN topologies provide broadcast filtering, security, address summarization, and traffic flow management. None of the switches within the defined group bridge any frames, not even broadcast frames, between two VLANs. Several key issues need to be considered when designing and building switched LAN internetworks.

- LAN segmentation
- Security
- Broadcast control
- Performance
- Network management
- Communication between VLANs

LAN Segmentation

VLANs allow logical network topologies to overlay the physical, switched infrastructure such that any arbitrary collection of LAN ports can be combined into an autonomous user group or community of interest. The technology logically segments the network into separate Layer 2 broadcast domains whereby packets are switched between ports designated to be within the same VLAN. By containing traffic originating on a particular LAN sent only to other LANs in the same VLAN, switched virtual networks avoid wasting bandwidth—a drawback inherent to traditional bridged and switched networks in which packets are often forwarded to LANs with no need for them. Implementation of VLANs also improves scalability, particularly in LAN environments that support broadcast- or multicast-intensive protocols and applications that flood packets throughout the network.

Figure 17-1 illustrates the difference between traditional physical LAN segmentation and logical VLAN segmentation.

Figure 17-1 *LAN Segmentation and VLAN Segmentation*

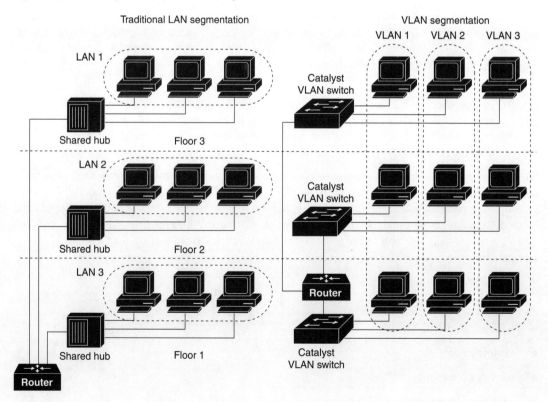

Security

VLANs also improve security by isolating groups. High-security users can be grouped into a VLAN, possibly on the same physical segment, and no users outside that VLAN can communicate with them.

Broadcast Control

Just as switches isolate collision domains for attached hosts and forward appropriate traffic through a particular port, VLANs provide complete isolation between VLANs. A VLAN is a bridging domain and all broadcast and multicast traffic is contained within it.

Performance

The logical grouping of users allows an accounting group to make intensive use of a networked accounting system assigned to a VLAN that contains just that accounting group and its servers. That group's work does not affect other users. The VLAN configuration improves general network performance by not slowing down other users sharing the network.

Network Management

The logical grouping of users allows easier network management. It is not necessary to pull cables to move a user from one network to another. Adds, moves, and changes are achieved by configuring a port into the appropriate VLAN.

Communication Between VLANs

Communication between VLANs is accomplished through routing, and the traditional security and filtering functions of the router can be used. Cisco IOS software provides network services such as security filtering, QoS, and accounting on a per-VLAN basis. As switched networks evolve to distributed VLANs, Cisco IOS provides key inter-VLAN communications and allows each network to scale.

VLAN Colors

VLAN switching is accomplished through *frame tagging,* where traffic originating and contained within a particular virtual topology carries a unique VLAN identifier (VLAN ID) as it traverses a common backbone or trunk link. The VLAN ID enables VLAN switching devices to make intelligent forwarding decisions based on the embedded VLAN ID. Each VLAN is differentiated by a *color,* or VLAN identifier. The unique VLAN ID determines the *frame coloring* for the VLAN. Packets originating and contained within a particular VLAN carry the identifier that uniquely defines that VLAN (by the VLAN ID).

The VLAN ID allows VLAN switches and routers to selectively forward packets to ports with the same VLAN ID. The switch that receives the frame from the source station inserts the VLAN ID and the packet is switched onto the shared backbone network. When the frame exits the switched LAN, a switch strips the header and forwards the frame to interfaces that match the VLAN color. If you are using a Cisco network management product such as VlanDirector, you can actually color-code the VLANs and monitor VLAN graphically.

Why Implement VLANs?

Network managers can logically group networks that span all major topologies, including high-speed technologies such as ATM, FDDI, and Fast Ethernet. By creating VLANs, system and network administrators can control traffic patterns, react quickly to relocations, and keep up with constant changes in the network due to moving requirements and node relocation just by changing the VLAN member list in the router configuration. They can add, remove, or move devices or make other changes to network configuration by using software to make the changes.

Issues regarding benefits of creating VLANs should have been addressed when you developed your network design. Issues to consider include

- Scalability

- Performance improvements

- Security

- Network additions, moves, and changes

Communicating Between VLANs

Cisco IOS provides full-feature routing at Layer 3, and translation at Layer 2 between VLANs. There are three different protocols available for routing between VLANs:

- Inter-Switch Link (ISL)

- IEEE 802.10

- ATM LAN emulation

All three of these technologies are based on OSI Layer 2 bridge-multiplexing mechanisms.

ISL Protocol

ISL protocol is used to interconnect two VLAN-capable Fast Ethernet devices, such as the Catalyst 5000 or 3000 switches and Cisco 7500 routers. The ISL protocol is a packet-tagging protocol that contains a standard Ethernet frame and the VLAN information associated with that frame. The packets on the ISL link contain a standard Ethernet, FDDI, or Token Ring frame and the VLAN information associated with that frame. ISL is currently supported only over Fast Ethernet links, but a single ISL link, or trunk, can carry different protocols from multiple VLANs.

IEEE 802.10 Protocol

The IEEE 802.10 protocol provides connectivity between VLANs. Originally developed to address the growing need for security within shared LAN/MAN environments, it incorporates authentication and encryption techniques to ensure data confidentiality and integrity throughout the network. Additionally, by functioning at Layer 2, it is well suited to high-throughput, low-latency switching environments. IEEE 802.10 protocol can run over any LAN or HDLC serial interface.

ATM LANE Protocol

The ATM LAN Emulation (LANE) protocol provides a way for legacy LAN users to take advantage of ATM benefits without requiring modifications to end-station hardware or software. LANE emulates a broadcast environment like IEEE 802.3 Ethernet on top of an ATM network that is a point-to-point environment.

LANE makes ATM function like a LAN. LANE allows standard LAN drivers such as NDIS and ODI to be used. The VLAN is transparent to applications. Applications can use normal LAN functions without dealing with the underlying complexities of the ATM implementation. For example, a station can send broadcasts and multicasts, even though ATM is defined as a point-to-point technology and doesn't support any-to-any services.

To accomplish this, special low-level software is implemented on an ATM client workstation, called the LAN Emulation Client, or LEC. The client software communicates with a central control point called a LAN Emulation Server, or LES. A Broadcast and Unknown Server (BUS) acts as a central point to distribute broadcasts and multicasts. The LAN Emulation Configuration Server (LECS) holds a database of LECs and the ELANs they belong to. The database is maintained by a network administrator.

VLAN Interoperability

Cisco IOS features bring added benefits to the VLAN technology. Enhancements to ISL, IEEE 802.10, and ATM LANE implementations enable routing of all major protocols between VLANs. These enhancements allow users to create more robust networks incorporating VLAN configurations by providing communications capabilities between VLANs.

Inter-VLAN Communications

The Cisco IOS software supports full routing of several protocols over ISL and ATM LANE virtual LANs. IP, Novell IPX, and AppleTalk routing are supported over IEEE 802.10 VLANs. Standard routing attributes, such as network advertisements, secondaries, and help addresses are applicable and VLAN routing is fast switched. Table 17-1 shows protocols supported for each VLAN encapsulation format and corresponding Cisco IOS releases.

Table 17-1 *Inter-VLAN Routing Protocol Support*

Protocol	ISL	ATM LANE	IEEE 802.10
IP	Release 11.1	Release 10.3	Release 11.1
Novell IPX (default encapsulation)	Release 11.1	Release 10.3	Release 11.1
Novell IPX (configurable encapsulation)	Release 11.3	Release 10.3	Release 11.3
AppleTalk Phase II	Release 11.3	Release 10.3	
DECnet	Release 11.3	Release 11.0	
Banyan VINES	Release 11.3	Release 11.2	
XNS	Release 11.3	Release 11.2	

VLAN Translation

VLAN translation refers to the ability of the Cisco IOS software to translate between different virtual LANs or between VLAN and non-VLAN encapsulating interfaces at Layer 2. Translation is typically used for selective inter-VLAN switching of non-routable protocols and to extend a single VLAN topology across hybrid switching environments. It is also possible to bridge VLANs on the main interface; the VLAN encapsulating header is preserved. Topology changes in one VLAN domain do not affect a different VLAN.

Designing Switched VLANs

By the time you are ready to configure routing between VLANs, you will have already defined them through the switches in your network. Issues related to network design and VLAN definition should be addressed during your network design:

- Sharing resources between VLANs
- Load balancing
- Redundant links
- Addressing
- Segmenting networks with VLANs

Segmenting the network into broadcast groups improves network security. Use router access lists based on station addresses, application types, and protocol types.

- Routers and their role in switched networks

In switched networks, routers perform broadcast management and route processing and distribution, and they provide communications between VLANs. Routers provide VLAN access to shared resources and connect to other parts of the network that are either logically segmented with the more traditional subnet approach or require access to remote sites across wide-area links.

CHAPTER 18

Configuring Routing Between VLANs with ISL Encapsulation

This chapter describes the ISL protocol and provides guidelines for configuring ISL and Token Ring ISL (TRISL) features. To locate documentation of other commands that appear in this chapter, you can search online at www.cisco.com.

Overview of ISL Protocol

ISL is a Cisco protocol for interconnecting multiple switches and maintaining VLAN information as traffic goes between switches. ISL provides VLAN capabilities while maintaining full-wire speed performance on Fast Ethernet links in full- or half-duplex mode. ISL operates in a point-to-point environment and supports up to 1,000 VLANs. You can define as many logical networks as are necessary for your environment.

This chapter describes how to configure routing between VLANs using ISL encapsulation.

Frame Tagging in ISL

With ISL, an Ethernet frame is encapsulated with a header that transports VLAN IDs between switches and routers. A 26-byte header that contains a 10-bit VLAN ID is prepended to the Ethernet frame.

A VLAN ID is added to the frame only when the frame is destined for a nonlocal network. Figure 18-1 illustrates VLAN packets traversing the shared backbone. Each VLAN packet carries the VLAN ID within the packet header.

Figure 18-1 *VLAN Packets Traversing the Shared Backbone*

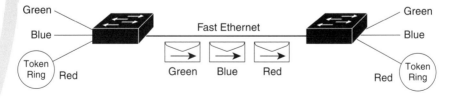

ISL Encapsulation Configuration Task List

You can configure routing between any number of VLANs in your network. This section documents the configuration tasks for each protocol supported with ISL encapsulation. The basic process is the same, regardless of the protocol being routed. It involves

- Enabling the protocol on the router.

- Enabling the protocol on the interface.

- Defining the encapsulation format as ISL or TRISL.

- Customizing the protocol according to the requirements for your environment.

The configuration processes documented in this chapter include the following:

- Configuring AppleTalk Routing over ISL

- Configuring Banyan VINES Routing over ISL

- Configuring DECnet Routing over ISL

- Configuring Hot Standby Router Protocol over ISL

- Configuring IP Routing over TRISL

- Configuring IPX Routing over TRISL

- Configuring VIP Distributed Switching over ISL

- Configuring XNS Routing over ISL

Refer to the "ISL Encapsulation Configuration Examples" section at the end of this chapter for sample configurations.

Configuring AppleTalk Routing over ISL

AppleTalk can be routed over VLAN subinterfaces by using the ISL and IEEE 802.10 VLAN encapsulation protocols. The AppleTalk routing over ISL and IEEE 802.10 VLANs feature provides full-feature Cisco IOS software AppleTalk support on a per-VLAN basis, allowing standard AppleTalk capabilities to be configured on VLANs.

To route AppleTalk over ISL or IEEE 802.10 between VLANs, customize the subinterface to create the environment in which it will be used by performing these tasks in the order in which they appear:

- Enabling AppleTalk Routing

- Defining the VLAN Encapsulation Format

- Configuring AppleTalk on the Subinterface

Enabling AppleTalk Routing

To enable AppleTalk routing on either ISL or 802.10 interfaces, use this command in global configuration mode:

Command	Purpose
appletalk routing [**eigrp** *router-number*]	Enables AppleTalk routing globally.

Defining the VLAN Encapsulation Format

To define the VLAN encapsulation format as either ISL or 802.10, use the following commands in interface configuration mode:

Step	Command	Purpose
1	**interface** *type slot/port.subinterface-number*	Specifies the subinterface the VLAN will use.
2	**encapsulation isl** *vlan-identifier* **encapsulation sde** *said*	Defines the encapsulation format as either ISL (**isl**) or IEEE 802.10 (**sde**), and specifies the VLAN identifier or security association identifier, respectively.

Configuring AppleTalk on the Subinterface

After you enable AppleTalk globally and define the encapsulation format, you need to enable it on the subinterface by specifying the cable range and naming the AppleTalk zone for each interface. To enable the AppleTalk protocol on the subinterface, use the following commands in interface configuration mode:

Command	Purpose
appletalk cable-range *cable-range* [*network.node*]	Assigns the AppleTalk cable range and zone for the subinterface.
appletalk zone *zone-name*	Assigns the AppleTalk zone for the subinterface.

Configuring Banyan VINES Routing over ISL

Banyan VINES can be routed over VLAN subinterfaces by using the ISL encapsulation protocol. The Banyan VINES routing over ISL VLANs feature provides full-feature Cisco IOS software Banyan VINES support on a per-VLAN basis, allowing standard Banyan VINES capabilities to be configured on VLANs.

To route Banyan VINES over ISL between VLANs, configure ISL encapsulation on the subinterface by performing these tasks in the order in which they appear:

- Enabling Banyan VINES Routing
- Defining the VLAN Encapsulation Format
- Configuring Banyan VINES on the Subinterface

Enabling Banyan VINES Routing

To begin the VINES routing configuration, use the following command in global configuration mode:

Command	Purpose
vines routing [*address*]	Enables Banyan VINES routing globally.

Defining the VLAN Encapsulation Format

To define the VINES routing encapsulation format, use the following commands in interface configuration mode:

Step	Command	Purpose
1	**interface** *type slot/port.subinterface-number*	Specifies the subinterface on which ISL will be used.
2	**encapsulation isl** *vlan-identifier*	Defines the encapsulation format as ISL (**isl**), and specifies the VLAN identifier.

Configuring Banyan VINES on the Subinterface

After you enable Banyan VINES globally and define the encapsulation format, you need to enable VINES on the subinterface by specifying the VINES routing metric. To enable the Banyan VINES protocol on the subinterface, use the following command in interface configuration mode:

Command	Purpose
vines metric [*whole* [*fractional*]]	Enables VINES routing on an interface.

Configuring DECnet Routing over ISL

DECnet can be routed over VLAN subinterfaces using the ISL VLAN encapsulation protocols. The DECnet routing over ISL VLANs feature provides full-feature Cisco IOS software DECnet support on a per-VLAN basis, allowing standard DECnet capabilities to be configured on VLANs.

To route DECnet over ISL VLAN, configure ISL encapsulation on the subinterface by performing these tasks in the order in which they appear:

● Enabling DECnet Routing

● Defining the VLAN Encapsulation Format

● Configuring DECnet on the Subinterface

Enabling DECnet Routing

To begin the DECnet routing configuration, use the following command in global configuration mode:

Command	Purpose
decnet [*network-number*] **routing** [*decnet-address*]	Enables DECnet on the router.

Defining the VLAN Encapsulation Format

To define the encapsulation format, use the following commands in interface configuration mode:

Step	Command	Purpose
1	**interface** *type slot/port.subinterface-number*	Specifies the subinterface on which ISL will be used.
2	**encapsulation isl** *vlan_identifier*	Defines the encapsulation format as ISL (**isl**), and specifies the VLAN identifier.

Configuring DECnet on the Subinterface

To configure DECnet routing on the subinterface, use the following command in interface configuration mode:

Command	Purpose
decnet cost [*cost-value*]	Enables DECnet routing on an interface.

Configuring Hot Standby Router Protocol over ISL

Hot Standby Router Protocol (HSRP) provides fault tolerance and enhanced routing performance for IP networks. HSRP allows Cisco IOS routers to monitor each other's operational status and very quickly assume packet forwarding responsibility in the event that the current forwarding device in the HSRP group fails or is taken down for maintenance. The standby mechanism remains transparent to the attached hosts and can be deployed on any LAN type. With multiple hot standby groups, routers can simultaneously provide redundant backup and perform load sharing across different IP subnets. Figure 18-2 illustrates HSRP in use with ISL providing routing between several VLANs.

Figure 18-2 *HSRP in VLAN Configurations*

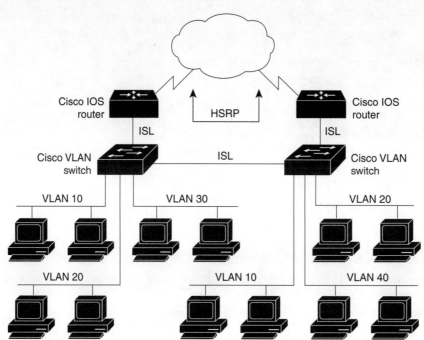

A separate HSRP group is configured for each VLAN subnet so that Cisco IOS router A can be the primary and forwarding router for VLANs 10 and 20. At the same time, it acts as backup for VLANs 30 and 40. Conversely, Router B acts as the primary and forwarding router for ISL VLANs 30 and 40, as well as the secondary and backup router for distributed VLAN subnets 10 and 20.

Running HSRP over ISL allows users to configure redundancy between multiple routers that are configured as front ends for VLAN IP subnets. By configuring HSRP over ISLs, users can eliminate situations in which a single point of failure causes traffic interruptions. This feature inherently provides some improvement in overall networking resilience by providing load balancing and redundancy capabilities between subnets and VLANs.

To configure HSRP over ISLs between VLANs, create the environment in which it will be used by performing these tasks in the order in which they appear:

● Defining the Encapsulation Format

● Defining the IP Address

● Enabling HSRP

Defining the Encapsulation Format

To define the encapsulation format as ISL, use the following commands in interface configuration mode:

Step	Command	Purpose
1	**interface** *type slot/port.subinterface-number*	Specifies the subinterface on which ISL will be used.
2	**encapsulation isl** *vlan-identifier*	Defines the encapsulation format, and specify the VLAN identifier.

Defining the IP Address

After you have specified the encapsulation format, define the IP address over which HSRP will be routed. Use the following command in interface configuration mode:

Command	Purpose
ip address *ip-address mask* [**secondary**]	Specifies the IP address for the subnet on which ISL will be used.

Enabling HSRP

To enable HSRP on an interface, enable the protocol, and then customize it for the interface. Use the following command in interface configuration mode:

Command	Purpose
standby [*group-number*] **ip** [*ip-address* [**secondary**]]	Enables HSRP.

To customize hot standby group attributes, use one or more of the following commands in interface configuration mode:

Command	Purpose
standby [*group-number*] **timers** *hellotime holdtime*	Configures the time between hello packets and the hold time before other routers declare the active router to be down.
standby [*group-number*] **priority** *priority*	Sets the hot standby priority used to choose the active router.
standby [*group-number*] **preempt**	Specifies that if the local router has priority over the current active router, the local router should attempt to take its place as the active router.

Command	Purpose
standby [*group-number*] **track** *type-number* [*interface-priority*]	Configures the interface to track other interfaces, so that if one of the other interfaces goes down, the hot standby priority for the device is lowered.
standby [*group-number*] **authentication** *string*	Selects an authentication string to be carried in all HSRP messages.

Configuring IP Routing over TRISL

The IP routing over TRISL VLANs feature extends IP routing capabilities to include support for routing IP frame types in VLAN configurations.

Enabling IP Routing

IP routing is automatically enabled in the Cisco IOS software for routers. To re-enable IP routing if it has been disabled, use the following command in global configuration mode:

Command	Purpose
ip routing	Enables IP routing on the router.

When you have IP routing enabled on the router, you can customize the characteristics to suit your environment.

Defining the VLAN Encapsulation Format

To define the encapsulation format as TRISL, use the following commands in interface configuration mode:

Step	Command	Purpose
1	**interface** *type slot/port.subinterface-number*	Specifies the subinterface on which TRISL will be used.
2	**encapsulation tr-isl trbrf-vlan** *vlanid* **bridge-num** *bridge-number*	Defines the encapsulation for TRISL.

The DRiP database is automatically enabled when TRISL encapsulation is configured, and at least one TrBRF is defined, and the interface is configured for SRB or for routing with RIF.

Assigning IP Address to Network Interface

An interface can have one primary IP address. To assign a primary IP address and a network mask to a network interface, use the following command in interface configuration mode:

Command	Purpose
ip address *ip-address mask*	Sets a primary IP address for an interface.

A mask identifies the bits that denote the network number in an IP address. When you use the mask to subnet a network, the mask is then referred to as a *subnet mask*.

NOTE   TRISL encapsulation must be specified for a subinterface before an IP address can be assigned to that subinterface.

Configuring IPX Routing over ISL

The IPX routing over ISL VLANs feature extends Novell NetWare routing capabilities to include support for routing all standard IPX encapsulations for Ethernet frame types in VLAN configurations. Users with Novell NetWare environments can now configure any one of the four IPX Ethernet encapsulations to be routed using the ISL encapsulation across VLAN boundaries. IPX encapsulation options now supported for VLAN traffic include

- *novell-ether* (Novell Ethernet_802.3)
- *sap* (Novell Ethernet_802.2)
- *arpa* (Novell Ethernet_II)
- *snap* (Novell Ethernet_Snap)

NetWare users can now configure consolidated VLAN routing over a single VLAN trunking interface. With configurable Ethernet encapsulation protocols, users have the flexibility of using VLANs regardless of their NetWare Ethernet encapsulation. Configuring Novell IPX encapsulations on a per-VLAN basis facilitates migration between versions of Netware. NetWare traffic can now be routed across VLAN boundaries with standard encapsulation options (*arpa*, *sap*, and *snap*) previously unavailable.

NOTE   Only one type of IPX encapsulation can be configured per VLAN (subinterface). The IPX encapsulation used must be the same within any particular subnet: A single encapsulation must be used by all NetWare systems that belong to the same virtual LAN.

To configure Cisco IOS software on a router with connected VLANs to exchange different IPX framing protocols, perform these tasks in the order in which they are appear:

● Enabling NetWare Routing

● Defining the VLAN Encapsulation Format

● Configuring NetWare on the Subinterface

Enabling NetWare Routing

To enable IPX routing on ISL interfaces, use the following command in global configuration mode:

Command	Purpose
ipx routing [*node*]	Enables IPX routing globally.

Defining the VLAN Encapsulation Format

To define the encapsulation format as ISL, use the following commands in interface configuration mode:

Step	Command	Purpose
1	**interface** *type slot/port.subinterface-number*	Specifies the subinterface on which ISL will be used.
2	**encapsulation isl** *vlan-identifier*	Defines the encapsulation format and specify the VLAN identifier.

Configuring NetWare on the Subinterface

After you enable NetWare globally and define the VLAN encapsulation format, you need to enable the subinterface by specifying the NetWare network number (if necessary) and the encapsulation type. Use this command in interface configuration mode:

Command	Purpose
ipx network *network* **encapsulation** *encapsulation-type*	Specifies the IPX encapsulation.

NOTE   The default IPX encapsulation format for Cisco IOS routers is *novell-ether* (Novell Ethernet_802.3). If you are running Novell Netware 3.12 or 4.0, the new Novell default encapsulation format is Novell Ethernet_802.2 and you should configure the Cisco router with the IPX encapsulation format *sap*.

Configuring IPX Routing over TRISL

The IPX routing over ISL VLANs feature extends Novell NetWare routing capabilities to include support for routing all standard IPX encapsulations for Ethernet frame types in VLAN configurations. Users with Novell NetWare environments can configure either *sap* or *snap* encapsulations to be routed using the TRISL encapsulation across VLAN boundaries. The *sap* (Novell Ethernet_802.2) IPX encapsulation is supported for VLAN traffic.

NetWare users can now configure consolidated VLAN routing over a single VLAN trunking interface. With configurable Ethernet encapsulation protocols, users have the flexibility of using VLANs regardless of their NetWare Ethernet encapsulation. Configuring Novell IPX encapsulations on a per-VLAN basis facilitates migration between versions of Netware. NetWare traffic can now be routed across VLAN boundaries with standard encapsulation options (*sap* and *snap*) previously unavailable.

NOTE	Only one type of IPX encapsulation can be configured per VLAN (subinterface). The IPX encapsulation used must be the same within any particular subnet: A single encapsulation must be used by all NetWare systems that belong to the same LANs.

To configure Cisco IOS software to exchange different IPX framing protocols on a router with connected VLANs, perform these tasks in the order in which they appear:

● Enabling NetWare Routing

● Defining the VLAN Encapsulation Format

● Configuring NetWare on the Subinterface

Enabling NetWare Routing

To enable IPX routing on TRISL interfaces, use the following command in global configuration mode:

Command	Purpose
ipx routing [*node*]	Enables IPX routing globally.

Defining the VLAN Encapsulation Format

To define the encapsulation format as TRISL, use the following commands in interface configuration mode:

Step	Command	Purpose
1	**interface** *type slot/port.subinterface-number*	Specifies the subinterface on which TRISL will be used.
2	**encapsulation tr-isl trbrf-vlan** *trbrf-vlan* **bridge-num** *bridge-num*	Defines the encapsulation for TRISL.

Configuring NetWare on the Subinterface

After you enable NetWare globally and define the VLAN encapsulation format, you need to enable the subinterface by specifying the NetWare network number (if necessary) and the encapsulation type. Use the following command in interface configuration mode:

Command	Purpose
ipx network *network* **encapsulation** *encapsulation-type*	Specifies the IPX encapsulation.

NOTE   The default IPX encapsulation format for Cisco IOS routers is *novell-ether* (Novell Ethernet_802.3). If you are running Novell Netware 3.12 or 4.0, the new Novell default encapsulation format is Novell Ethernet_802.2 and you should configure the Cisco router with the IPX encapsulation format *sap*.

Configuring VIP Distributed Switching over ISL

With the introduction of the VIP Distributed ISL feature, ISL encapsulated IP packets can be switched on VIP controllers installed on Cisco 7500 series routers.

The VIP2 provides distributed switching of IP encapsulated in ISL in VLAN configurations. Where an aggregation route performs inter-VLAN routing for multiple VLANs, traffic can be switched autonomously on-card or between cards rather than through the central RSP. Figure 18-3 shows the VIP distributed architecture of the Cisco 7500 series router.

Figure 18-3 *Cisco 7500 Distributed Architecture*

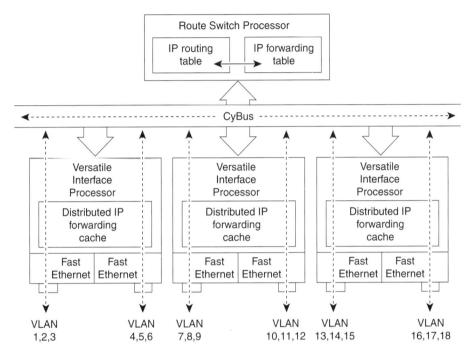

This distributed architecture allows incremental capacity increases by installing additional VIP cards. Using VIP cards for switching the majority of IP VLAN traffic in multiprotocol environments significantly increases routing performance for the other protocols since the RSP offloads IP and can then be dedicated to switching the non-IP protocols.

VIP distributed switching offloads switching of ISL VLAN IP traffic to the VIP card, removing involvement from the main CPU. Offloading ISL traffic to the VIP card significantly improves networking performance. Because you can install multiple VIP cards in a router, VLAN routing capacity is increased linearly according to the number of VIP cards installed in the router.

To configure distributed switching on the VIP, you must first configure the router for IP routing by performing these tasks in the order in which they appear:

● Enabling IP Routing

● Enabling VIP Distributed Switching

● Configuring ISL Encapsulation on the Subinterface

Enabling IP Routing

To enable IP routing, use the following command in global configuration mode:

Command	Purpose
ip routing	Enables IP routing on the router.

When you have IP routing enabled on the router, you can customize the characteristics to suit your environment.

Enabling VIP Distributed Switching

To enable VIP distributed switching, use the following commands beginning in interface configuration mode:

Step	Command	Purpose
1	**interface** *type slot/port-adapter/port*	Specifies the interface and enters interface configuration mode.
2	**ip route-cache distributed**	Enables VIP distributed switching of IP packets on the interface.

Configuring ISL Encapsulation on the Subinterface

To configure ISL encapsulation on the subinterface, use the following commands in interface configuration mode:

Step	Command	Purpose
1	**interface** *type slot/port-adapter/port*	Specifies the interface and enters interface configuration mode.
2	**encapsulation isl** *vlan-identifier*	Defines the encapsulation format as ISL and specifies the VLAN identifier.

Configuring XNS Routing over ISL

XNS can be routed over VLAN subinterfaces using the ISL VLAN encapsulation protocol. The XNS routing over ISL VLANs feature provides full-feature Cisco IOS software XNS support on a per-VLAN basis, allowing standard XNS capabilities to be configured on VLANs.

To route XNS over ISL VLANs, configure ISL encapsulation on the subinterface by performing these tasks in the order in which they appear:

- Enabling XNS Routing

- Defining the VLAN Encapsulation Format

- Configuring XNS on the Subinterface

Enabling XNS Routing

Begin the XNS routing configuration by using the following commands in global configuration mode:

Command	Purpose
xns routing [*address*]	Enables XNS routing globally.

Defining the VLAN Encapsulation Format

To define the VLAN encapsulation format, use the following commands in interface configuration mode:

Step	Command	Purpose
1	**interface** *type slot/port.subinterface-number*	Specifies the subinterface on which ISL will be used.
2	**encapsulation isl** *vlan-identifier*	Defines the encapsulation format as ISL (**isl**, and specifies the VLAN identifier.

Configuring XNS on the Subinterface

Enable XNS on the subinterface by specifying the XNS network number. Use the following command in interface configuration mode:

Command	Purpose
xns network [*number*]	Enables XNS routing on the subinterface.

ISL Encapsulation Configuration Examples

This section provides configuration examples for each of the protocols described in this chapter. It includes these examples:

- AppleTalk routing over ISL configuration example

- Banyan VINES routing over ISL configuration example

- DECnet routing over ISL configuration example

- HSRP over ISL configuration example

- IP routing over RIF between TrBRF VLANs

- IP routing between a TRISL VLAN and an Ethernet ISL VLAN

- IPX routing over ISL configuration example

- Routing with RIF between a TRISL VLAN and a Token Ring interface

- VIP distributed switching over ISL configuration example

- XNS routing over ISL configuration example

AppleTalk Routing over ISL Configuration Example

The configuration example illustrated in Figure 18-4 shows AppleTalk being routed between different ISL and IEEE 802.10 VLAN encapsulating subinterfaces.

Figure 18-4 *Routing AppleTalk over VLAN encapsulations*

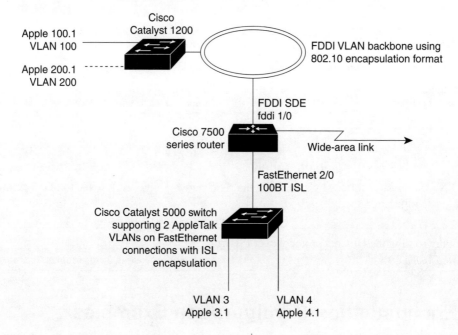

As shown in Figure 18-4, AppleTalk traffic is routed to and from switched VLAN domains 3, 4, 100, and 200 to any other AppleTalk routing interface. This example shows a sample configuration file for the Cisco 7500 series router with the commands entered to configure the network shown in Figure 18-4:

```
!
appletalk routing
interface Fddi 1/0.100
 encapsulation sde 100
 appletalk cable-range 100-100 100.2
 appletalk zone 100
!
interface Fddi 1/0.200
 encapsulation sde 200
 appletalk cable-range 200-200 200.2
 appletalk zone 200
!
interface FastEthernet 2/0.3
 encapsulation isl 3
 appletalk cable-range 3-3 3.2
 appletalk zone 3
!
interface FastEthernet 2/0.4
 encapsulation isl 4
 appletalk cable-range 4-4 4.2
 appletalk zone 4
!
```

Banyan VINES Routing over ISL Configuration Example

To configure routing of the Banyan VINES protocol over ISL trunks, you need to define ISL as the encapsulation type. This example shows Banyan VINES configured to be routed over an ISL trunk:

```
vines routing
interface fastethernet 0.1
 encapsulation isl 100
 vines metric 2
```

DECnet Routing over ISL Configuration Example

To configure routing the DECnet protocol over ISL trunks, you need to define ISL as the encapsulation type. This example shows DECnet configured to be routed over an ISL trunk:

```
decnet routing 2.1
interface fastethernet 1/0.1
 encapsulation isl 200
 decnet cost 4
```

HSRP over ISL Configuration Example

The configuration example shown in Figure 18-5 shows HSRP being used on two VLAN routers sending traffic to and from ISL VLANs through a Catalyst 5000 switch. Each router forwards its own traffic and acts as a standby for the other.

Figure 18-5 *HSRP Sample Configuration*

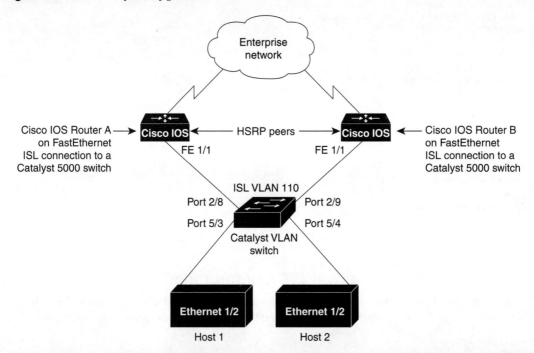

The topology shown in Figure 18-5 illustrates a Cisco Catalyst VLAN switch supporting Fast Ethernet connections to two routers running HSRP. Both routers are configured to route HSRP over ISLs.

The standby conditions are determined by the standby commands used in the configuration. Traffic from Host 1 is forwarded through Router A. Because the priority for the group is higher, Router A is the active router for Host 1. Because the priority for the group serviced by Host 2 is higher in Router B, traffic from Host 2 is forwarded through Router B, making Router B its active router.

In the configuration shown in Figure 18-5, if the active router becomes unavailable, the standby router assumes active status for the additional traffic and automatically routes the traffic normally handled by the router that has become unavailable.

The following is the Host 1 configuration:

```
interface Ethernet 1/2
 ip address 110.1.1.25 255.255.255.0
 ip route 0.0.0.0 0.0.0.0 110.1.1.101
```

The following is the Host 2 configuration:

```
interface Ethernet 1/2
 ip address 110.1.1.27 255.255.255.0
 ip route 0.0.0.0 0.0.0.0 110.1.1.102
!
```

The following is the Router A configuration:

```
interface FastEthernet 1/1.110
 encapsulation isl 110
 ip address 110.1.1.2 255.255.255.0
 standby 1 ip 110.1.1.101
 standby 1 preempt
 standby 1 priority 105
 standby 2 ip 110.1.1.102
 standby 2 preempt

!
end

!
```

The following is the Router B configuration:

```
interface FastEthernet 1/1.110
 encapsulation isl 110
 ip address 110.1.1.3 255.255.255.0
 standby 1 ip 110.1.1.101
 standby 1 preempt
 standby 2 ip 110.1.1.102
 standby 2 preempt
 standby 2 priority 105
router igrp 1
!
network 110.1.0.0
network 120.1.0.0
!
```

The following is the VLAN switch configuration:

```
set vlan 110 5/4
set vlan 110 5/3
set trunk 2/8 110
set trunk 2/9 110
```

IP Routing with RIF Between TrBRF VLANs

Figure 18-6 illustrates IP routing with RIF between two TrBRF VLANs.

Figure 18-6 *IP Routing with RIF Between TrBRF VLANs*

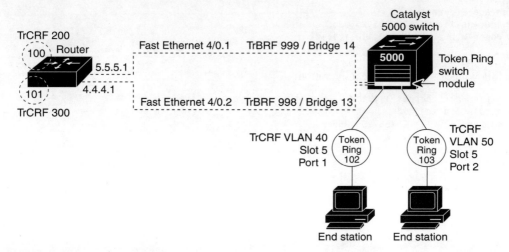

The following is the configuration for the router:

```
interface FastEthernet4/0.1
  ip address 5.5.5.1 255.255.255.0
  encapsulation tr-isl trbrf-vlan 999 bridge-num 14
  multiring trcrf-vlan 200 ring 100
  multiring all
!
  interface FastEthernet4/0.2
  ip address 4.4.4.1 255.255.255.0
  encapsulation tr-isl trbrf-vlan 998 bridge-num 13
  multiring trcrf-vlan 300 ring 101
  multiring all
```

The following is the configuration for the Catalyst 5000 switch with the Token Ring switch module in slot 5. In this configuration, the Token Ring port 102 is assigned with TrCRF VLAN 40 and the Token Ring port 103 is assigned with TrCRF VLAN 50:

```
#vtp
set vtp domain trisl
set vtp mode server
set vtp v2 enable
#drip
set set tokenring reduction enable
set tokenring distrib-crf disable
#vlans
set vlan 999 name trbrf type trbrf bridge 0xe stp ieee
set vlan 200 name trcrf200 type trcrf parent 999 ring 0x64 mode srb
set vlan 40 name trcrf40 type trcrf parent 999 ring 0x66 mode srb
set vlan 998 name trbrf type trbrf bridge 0xd stp ieee
set vlan 300 name trcrf300 type trcrf parent 998 ring 0x65 mode srb
set vlan 50 name trcrf50 type trcrf parent 998 ring 0x67 mode srb
#add token port to trcrf 40
```

```
set vlan 40   5/1
#add token port to trcrf 50
set vlan 50   5/2
set trunk 1/2 on
```

IP Routing Between a TRISL VLAN and an Ethernet ISL VLAN

Figure 18-7 illustrates IP routing between a TRISL VLAN and an Ethernet ISL VLAN.

Figure 18-7 *IP Routing Between a TRISL VLAN and an Ethernet ISL VLAN*

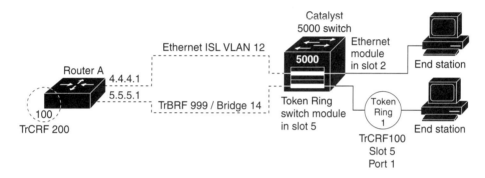

The following is the configuration for the router:

```
interface FastEthernet4/0.1
 ip address 5.5.5.1 255.255.255.0
 encapsulation tr-isl trbrf-vlan 999 bridge-num 14
 multiring trcrf-vlan 20 ring 100
 multiring all
!
interface FastEthernet4/0.2
 ip address 4.4.4.1 255.255.255.0
 encapsulation isl 12
```

IPX Routing over ISL Configuration Example

Figure 18-8 shows IPX interior encapsulations configured over ISL encapsulation in VLAN configurations. Note that three different IPX encapsulation formats are used. VLAN 20 uses *sap* encapsulation, VLAN 30 uses *arpa*, and VLAN 70 uses *novell-ether*. Prior to the introduction of this feature, only the default encapsulation format, *novell-ether*, was available for routing IPX over ISL links in VLANs.

Figure 18-8 *Configurable IPX Encapsulations Routed over ISL in VLAN Configurations*

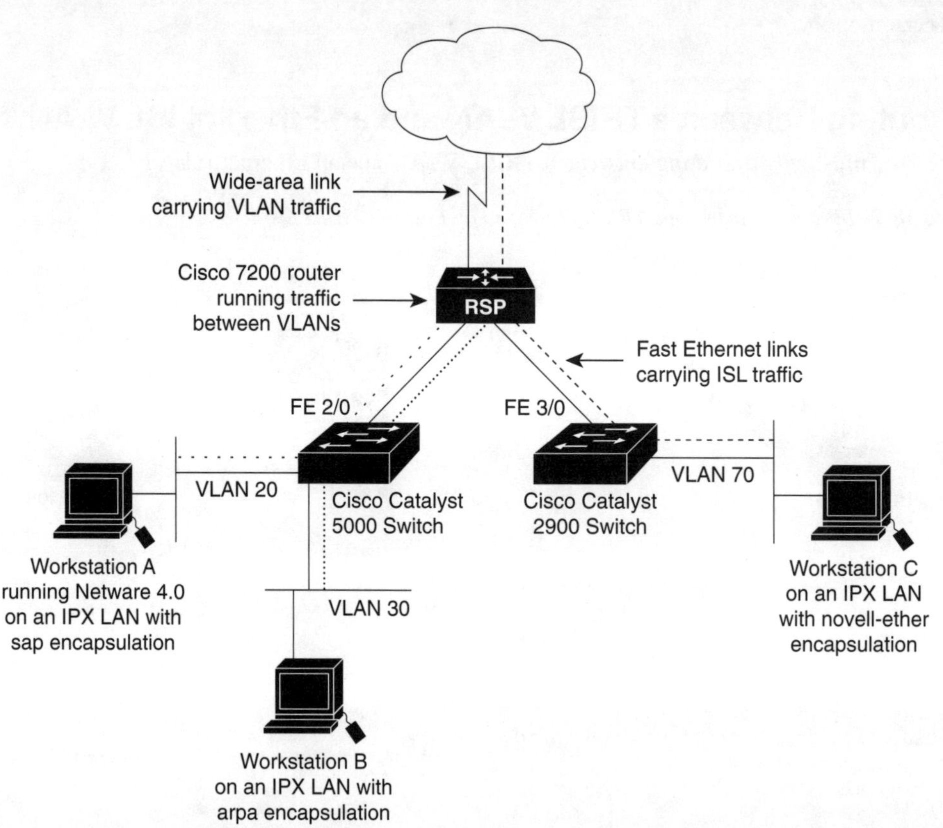

The following is the VLAN 20 configuration:

```
ipx routing
interface FastEthernet 2/0
 no shutdown
interface FastEthernet 2/0.20
 encapsulation isl 20
 ipx network 20 encapsulation sap
```

The following is the VLAN 30 configuration:

```
ipx routing
interface FastEthernet 2/0
 no shutdown
interface FastEthernet 2/0.30
 encapsulation isl 30
 ipx network 30 encapsulation arpa
```

The following is the VLAN 70 configuration:

```
ipx routing
interface FastEthernet 3/0
 no shutdown
interface Fast3/0.70
 encapsulation isl 70
 ipx network 70 encapsulation novell-ether
```

Routing with RIF Between a TRISL VLAN and a Token Ring Interface

Figure 18-9 illustrates routing with RIF between a TRISL VLAN and a Token Ring interface.

Figure 18-9 *Routing with RIF Between a TRISL VLAN and a Token Ring Interface*

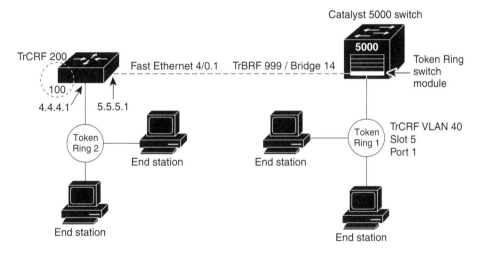

The following is the configuration for the router:

```
source-bridge ring-group 100
!
interface TokenRing 3/1
  ip address 4.4.4.1 255.255.255.0
!
interface FastEthernet4/0.1
  ip address 5.5.5.1 255.255.255.0
  encapsulation tr-isl trbrf 999 bridge-num 14
  multiring trcrf-vlan 200 ring-group 100
  multiring all
```

The following is the configuration for the Catalyst 5000 switch with the Token Ring switch module in slot 5. In this configuration, the Token Ring port 1 is assigned to the TrCRF VLAN 40:

```
#vtp
set vtp domain trisl
set vtp mode server
set vtp v2 enable
#drip
set set tokenring reduction enable
set tokenring distrib-crf disable
#vlans
set vlan 999 name trbrf type trbrf bridge 0xe stp ieee
set vlan 200 name trcrf200 type trcrf parent 999 ring 0x64 mode srt
set vlan 40 name trcrf40 type trcrf parent 999 ring 0x1 mode srt
#add token port to trcrf 40
set vlan 40    5/1
set trunk 1/2 on
```

VIP Distributed Switching over ISL Configuration Example

Figure 18-10 illustrates a topology in which Catalyst VLAN switches are connected to routers
forwarding traffic from a number of ISL VLANs. With the VIP distributed ISL capability in the
Cisco 7500 series router, each VIP card can route ISL-encapsulated VLAN IP traffic. The inter-VLAN
routing capacity is increased linearly by the packet-forwarding capability of each VIP card.

Figure 18-10 *VIP Distributed ISL VLAN Traffic*

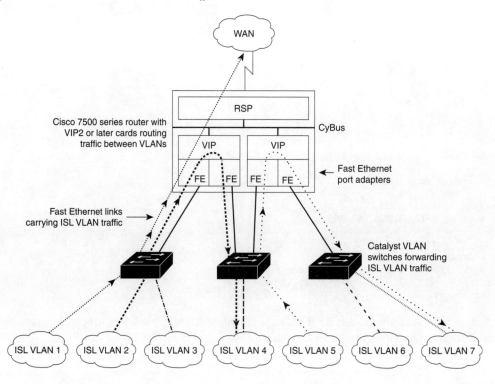

In Figure 18-10, the VIP cards forward the traffic between ISL VLANs or any other routing interface. Traffic from any VLAN can be routed to any of the other VLANs, regardless of which VIP card receives the traffic.

These commands show the configuration for each of the VLANs shown in Figure 18-10:

```
interface FastEthernet1/0/0
 ip address 20.1.1.1 255.255.255.0
 ip route-cache distributed
 full-duplex

interface FastEthernet1/0/0.1
 ip address 22.1.1.1 255.255.255.0
 encapsulation isl 1

interface FastEthernet1/0/0.2
 ip address 22.1.2.1 255.255.255.0
 encapsulation isl 2

interface FastEthernet1/0/0.3
 ip address 22.1.3.1 255.255.255.0
 encapsulation isl 3

interface FastEthernet1/1/0
 ip route-cache distributed
 full-duplex

interface FastEthernet1/1/0.1
 ip address 77.1.1.1 255.255.255.0
 encapsulation isl 4

interface FastEthernet 2/0/0
 ip address 30.1.1.1 255.255.255.0
 ip route-cache distributed
 full-duplex

interface FastEthernet2/0/0.5
 ip address 33.1.1.1 255.255.255.0
 encapsulation isl 5

interface FastEthernet2/1/0
 ip address 40.1.1.1 255.255.255.0
 ip route-cache distributed
 full-duplex

interface FastEthernet2/1/0.6
 ip address 44.1.6.1 255.255.255.0
 encapsulation isl 6

interface FastEthernet2/1/0.7
 ip address 44.1.7.1 255.255.255.0
 encapsulation isl 7
```

XNS Routing over ISL Configuration Example

To configure routing of the XNS protocol over ISL trunks, you need to define ISL as the encapsulation type. This example shows XNS configured to be routed over an ISL trunk:

```
xns routing 0123.4567.adcb
interface fastethernet 1/0.1
 encapsulation isl 100
 xns network 20
```

CHAPTER 19

Configuring Routing Between VLANs with IEEE 802.10 Encapsulation

This chapter describes the required and optional tasks for configuring routing between VLANs with IEEE 802.10 encapsulation. For a complete description of VLAN commands used in this chapter, refer to Chapter 3, "Cisco IOS Switching Commands." For documentation of other commands that appear in this chapter, you can search online at www.cisco.com.

The IEEE 802.10 standard provides a method for secure bridging of data across a shared backbone. It defines a single frame type known as the Secure Data Exchange (SDE), a MAC-layer frame with an IEEE 802.10 header inserted between the MAC header and the frame data. A well-known Logical Link Control Service Access Point notifies the switch of an incoming IEEE 802.10 frame. The VLAN ID is carried in the 4-byte Security Association Identifier (SAID) field.

HDLC serial links can be used as VLAN trunks in IEEE 802.10 virtual LANs to extend a virtual topology beyond a LAN backbone.

Configure AppleTalk Routing over IEEE 802.10

AppleTalk can be routed over VLAN subinterfaces by using the ISL or IEEE 802.10 VLAN encapsulation protocols. The AppleTalk Routing over IEEE 802.10 Virtual LANs feature provides full-feature Cisco IOS software AppleTalk support on a per-VLAN basis, allowing standard AppleTalk capabilities to be configured on VLANs.

AppleTalk users can now configure consolidated VLAN routing over a single VLAN trunking interface. Prior to introduction of this feature, AppleTalk could be routed only on the main interface on a LAN port. If AppleTalk routing was disabled on the main interface or if the main interface was shut down, the entire physical interface would stop routing any AppleTalk packets. With this feature enabled, AppleTalk routing on subinterfaces will be unaffected by changes in the main interface with the main interface in the "no-shut" state.

To route AppleTalk over IEEE 802.10 between VLANs, create the environment in which it will be used by customizing the subinterface by performing these tasks in the order in which they appear:

● Enabling AppleTalk Routing

● Configuring AppleTalk on the Subinterface

● Defining the VLAN Encapsulation Format

Enabling AppleTalk Routing

To enable AppleTalk routing on IEEE 802.10 interfaces, use the following command in global configuration mode:

Command	Purpose
appletalk routing [**eigrp** *router-number*]	Enables AppleTalk routing globally.

Configuring AppleTalk on the Subinterface

After you enable AppleTalk globally and define the encapsulation format, you need to enable it on the subinterface by specifying the cable range and naming the AppleTalk zone for each interface. To enable the AppleTalk protocol on the subinterface, use the following commands in interface configuration mode:

Command	Purpose
appletalk cable-range *cable-range* [*network.node*]	Assigns the AppleTalk cable range and zone for the subinterface.
appletalk zone *zone-name*	Assigns the AppleTalk zone for the subinterface.

Defining the VLAN Encapsulation Format

To define the VLAN encapsulation format as either ISL or 802.10, use the following commands in interface configuration mode:

Step	Command	Purpose
1	**interface** *type slot/port.subinterface-number*	Specifies the subinterface the VLAN will use.
2	**encapsulation sde** *said*	Defines the encapsulation format as IEEE 802.10 (**sde**) and specifies the VLAN identifier or security association identifier, respectively.

Routing AppleTalk over IEEE 802.10 Configuration Example

The configuration example illustrated in Figure 19-1 shows AppleTalk being routed between different ISL and IEEE 802.10 VLAN encapsulating subinterfaces.

Figure 19-1 *Routing AppleTalk over VLAN Encapsulations*

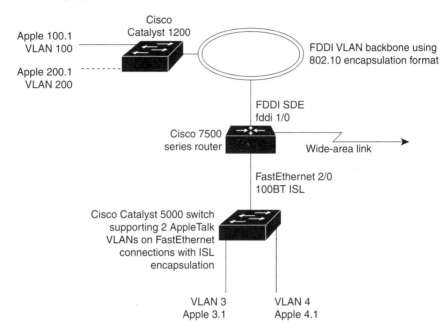

As shown in Figure 19-1, AppleTalk traffic is routed to and from switched VLAN domains 3, 4, 100, and 200 to any other AppleTalk routing interface. This example shows a sample configuration file for the Cisco 7500 series router with the commands entered to configure the network shown in Figure 19-1:

```
!
interface Fddi 1/0.100
 encapsulation sde 100
 appletalk cable-range 100-100 100.2
 appletalk zone 100
!
interface Fddi 1/0.200
 encapsulation sde 200
 appletalk cable-range 200-200 200.2
 appletalk zone 200
!
interface FastEthernet 2/0.3
 encapsulation isl 3
 appletalk cable-range 3-3 3.2
 appletalk zone 3
!
interface FastEthernet 2/0.4
 encapsulation isl 4
 appletalk cable-range 4-4 4.2
 appletalk zone 4
!
```

CHAPTER 20

LAN Emulation Overview

This overview chapter gives a high-level description of LANE. For specific configuration information, refer to Chapter 21, "Configuring LAN Emulation."

LANE

Cisco's implementation of LANE makes an ATM interface look like one or more Ethernet interfaces.

LANE is an ATM service defined by the ATM Forum specification *LAN Emulation over ATM*, ATM_FORUM 94-0035. This service emulates the following LAN-specific characteristics:

- Connectionless services

- Multicast services

- LAN MAC driver services

LANE service provides connectivity between ATM-attached devices and connectivity with LAN-attached devices. This includes connectivity between ATM-attached stations and LAN-attached stations and also connectivity between LAN-attached stations across an ATM network.

Because LANE connectivity is defined at the MAC layer, upper protocol layer functions of LAN applications can continue unchanged when the devices join emulated LANs. This feature protects corporate investments in legacy LAN applications.

An ATM network can support multiple independent emulated LAN networks. Membership of an end system in any of the emulated LANs is independent of the physical location of the end system. This characteristic enables easy hardware moves and location changes. In addition, the end systems can also move easily from one emulated LAN to another, whether or not the hardware moves.

LAN emulation in an ATM environment provides routing between emulated LANs for supported routing protocols and high-speed, scalable switching of local traffic.

The ATM LANE system has three servers that are single points of failure. These are the LECS, the LES, and the BUS. Beginning with Release 11.2, LANE fault tolerance or Simple LANE Service Replication on the emulated LAN provides backup servers to prevent problems if these servers fail.

The fault tolerance mechanism that eliminates these single points of failure is described in Chapter 21, "Configuring LAN Emulation." Although this scheme is proprietary, no new protocol additions have been made to the LANE subsystems.

LANE Components

Any number of emulated LANs can be set up in an ATM switch cloud. A router can participate in any number of these emulated LANs.

LANE is defined on a LAN client/server model. The following components are implemented in this release:

- LANE client

  A LANE client emulates a LAN interface to higher layer protocols and applications. It forwards data to other LANE components and performs LANE address resolution functions.

  Each LANE client is a member of only one emulated LAN. However, a router can include LANE clients for multiple emulated LANs: one LANE client for *each* emulated LAN of which it is a member.

  If a router has clients for multiple emulated LANs, the Cisco IOS software can route traffic between the emulated LANs.

- LANE server

  The LANE server for an emulated LAN is the control center. It provides joining, address resolution, and address registration services to the LANE clients in that emulated LAN. Clients can register destination unicast and multicast MAC addresses with the LANE server. The LANE server also handles LANE ARP (LE ARP) requests and responses.

  Our implementation has a limit of one LANE server per emulated LAN.

- LANE BUS

  The LANE BUS sequences and distributes multicast and broadcast packets and handles unicast flooding.

  In this release, the LANE server and the LANE BUS are combined and located in the same Cisco 7000 family or Cisco 4500 series router; one combined LANE server and BUS is required per emulated LAN.

- LANE configuration server

  The LANE configuration server contains the database that determines which emulated LAN a device belongs to (each configuration server can have a different named database). Each LANE client consults the LANE configuration server just once, when it joins an emulated LAN, to determine which emulated LAN it should join. The LANE configuration server returns the ATM address of the LANE server for that emulated LAN.

  One LANE configuration server is required per LANE ATM switch cloud.

The LANE configuration server's database can have the following four types of entries:

— Emulated LAN name-ATM address of LANE server pairs

— LANE client MAC address-emulated LAN name pairs

— LANE client ATM template-emulated LAN name pairs

— Default emulated LAN name

NOTE   Emulated LAN names must be unique on an interface. If two interfaces participate in
LANE, the second interface may be in a different switch cloud.

LANE Operation and Communication

Communication among LANE components is ordinarily handled by several types of switched virtual
circuits (SVCs). Some SVCs are unidirectional; others are bidirectional. Some are point-to-point and
others are point-to-multipoint. Figure 20-1 illustrates the various virtual channel connections (VCCs)—
also known as *virtual circuit connections*—that are used in LANE configuration.

Figure 20-1 *LANE VCC Types*

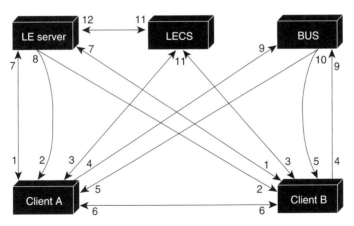

1–7	Control direct	4–9	Multicast send
2–8	Control distribute	5–10	Multicast forward
3–11	Configure direct (client)	6–6	Data direct
		11–12	Configure direct (server)

The following section describes various processes that occur, starting with a client requesting to join an
emulated LAN after the component routers have been configured.

Client Joining an Emulated LAN

The following process normally occurs after a LANE client has been enabled:

- Client requests to join an emulated LAN

  The client sets up a connection to the LANE configuration server—a bidirectional point-to-point Configure Direct VCC—to find the ATM address of the LANE server for its emulated LAN.

  LANE clients find the LANE configuration server by using the following methods in the listed order:

  — Locally configured ATM address

  — Interim Local Management Interface (ILMI)

  — Fixed address defined by the ATM Forum

  — PVC 0/17

- Configuration server identifies the LANE server

  Using the same VCC, the LANE configuration server returns the ATM address and the name of the LANE server for the client's emulated LAN.

- Client contacts the server for its LAN

  The client sets up a connection to the LANE server for its emulated LAN (a bidirectional point-to-point Control Direct VCC) to exchange control traffic.

  Once a Control Direct VCC is established between a LANE client and a LANE server, it remains up.

- Server verifies that the client is allowed to join the emulated LAN

  The server for the emulated LAN sets up a connection to the LANE configuration server to verify that the client is allowed to join the emulated LAN—a bidirectional point-to-point Configure Direct (server) VCC. The server's configuration request contains the client's MAC address, its ATM address, and the name of the emulated LAN. The LANE configuration server checks its database to determine whether the client can join that LAN; then it uses the same VCC to inform the server whether the client is or is not allowed to join.

- LANE server allows or disallows the client to join the emulated LAN

  If allowed, the LANE server adds the LANE client to the unidirectional point-to-multipoint Control Distribute VCC and confirms the join over the bidirectional point-to-point Control Direct VCC. If disallowed, the LANE server rejects the join over the bidirectional point-to-point Control Direct VCC.

- LANE client sends LE ARP packets for the broadcast address, which is all 1s

  Sending LE ARP packets for the broadcast address sets up the VCCs to and from the BUS.

Address Resolution

As communication occurs on the emulated LAN, each client dynamically builds a local LANE ARP (LE ARP) table. A client's LE ARP table can also have static, preconfigured entries. The LE ARP table maps MAC addresses to ATM addresses.

NOTE   LE ARP is not the same as IP ARP. IP ARP maps IP addresses (Layer 3) to Ethernet MAC addresses (Layer 2); LE ARP maps emulated LAN MAC addresses (Layer 2) to ATM addresses (also Layer 2).

When a client first joins an emulated LAN, its LE ARP table has no dynamic entries and the client has no information about destinations on or behind its emulated LAN. To learn about a destination when a packet is to be sent, the client begins the following process to find the ATM address corresponding to the known MAC address:

- The client sends an LE ARP request to the LANE server for this emulated LAN (point-to-point Control Direct VCC).

- The LANE server forwards the LE ARP request to all clients on the emulated LAN (point-to-multipoint Control Distribute VCC).

- Any client that recognizes the MAC address responds with its ATM address (point-to-point Control Direct VCC).

- The LANE server forwards the response (point-to-multipoint Control Distribute VCC).

- The client adds the MAC address-ATM address pair to its LE ARP cache.

- Then the client can establish a VCC to the desired destination and transmit packets to that ATM address (bidirectional point-to-point Data Direct VCC).

For unknown destinations, the client sends a packet to the BUS, which forwards the packet to all clients via flooding. The BUS floods the packet because the destination might be behind a bridge that has not yet learned this particular address.

Multicast Traffic

When a LANE client has broadcast or multicast traffic, or unicast traffic with an unknown address to send, the following process occurs:

- The client sends the packet to the BUS (unidirectional point-to-point Multicast Send VCC).

- The BUS forwards (floods) the packet to all clients (unidirectional point-to-multipoint Multicast Forward VCC).

This VCC branches at each ATM switch. The switch forwards such packets to multiple outputs. (The switch does not examine the MAC addresses; it simply forwards all packets it receives.)

Typical LANE Scenarios

In typical LANE cases, one or more Cisco 7000 family routers, or Cisco 4500 series routers are attached to a Cisco LightStream ATM switch. The LightStream ATM switch provides connectivity to the broader ATM network switch cloud. The routers are configured to support one or more emulated LANs. One of the routers is configured to perform the LANE configuration server functions. A router is configured to perform the server function and the BUS function for each emulated LAN. (One router can perform the server function and the BUS function for several emulated LANs.) In addition to these functions, each router also acts as a LANE client for one or more emulated LANs.

This section presents two scenarios using the same four Cisco routers and the same Cisco LightStream ATM switch. Figure 20-2 illustrates a scenario in which one emulated LAN is set up on the switch and routers. Figure 20-3 illustrates a scenario in which several emulated LANs are set up on the switch and routers.

The physical layout and the physical components of an emulated network might not differ for the single and the multiple emulated LAN cases. The differences are in the software configuration for the number of emulated LANs and the assignment of LANE components to the different physical components.

Single Emulated LAN Scenario

In a single emulated LAN scenario, the LANE components might be assigned as follows:

- Router 1 includes the following LANE components:
  - The LANE configuration server (one per LANE switch cloud)
  - The LANE server and BUS for the emulated LAN with the default name *man* (for manufacturing)
  - The LANE client for the *man* emulated LAN.
- Router 2 includes a LANE client for the *man* emulated LAN.
- Router 3 includes a LANE client for the *man* emulated LAN.
- Router 4 includes a LANE client for the *man* emulated LAN.

Figure 20-2 illustrates this single emulated LAN configured across several routers.

Figure 20-2 *Single Emulated LAN Configured on Several Routers*

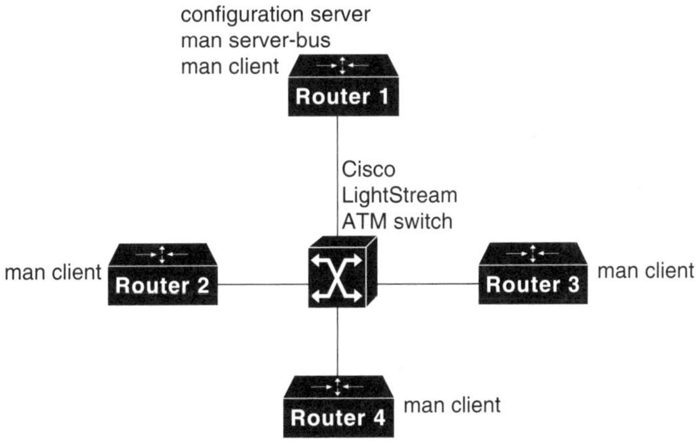

configuration server
man server-bus
man client

Router 1

Cisco
LightStream
ATM switch

man client Router 2

Router 3 man client

man client Router 4

Multiple Emulated LAN Scenario

In the multiple LAN scenario, the same switch and routers are used, but multiple emulated LANs are configured. See Figure 20-3.

Figure 20-3 *Multiple Emulated LANs Configured on Several Routers*

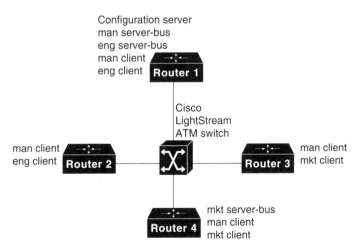

Configuration server
man server-bus
eng server-bus
man client
eng client

Router 1

Cisco
LightStream
ATM switch

man client
eng client Router 2

Router 3 man client
mkt client

mkt server-bus
Router 4 man client
mkt client

In the following scenario, three emulated LANs are configured on four routers:

- Router 1 includes the following LANE components:
    - The LANE configuration server (one per LANE switch cloud)
    - The LANE server and BUS for the emulated LAN called *man* (for manufacturing)
    - The LANE server and BUS functions for the emulated LAN called *eng* (for engineering)
    - A LANE client for the *man* emulated LAN
    - A LANE client for the *eng* emulated LAN
- Router 2 includes only the LANE clients for the *man* and *eng* emulated LANs.
- Router 3 includes only the LANE clients for the *man* and *mkt* (for marketing) emulated LANs.
- Router 4 includes the following LANE components:
    - The LANE server and BUS for the *mkt* emulated LAN
    - A LANE client for the *man* emulated LAN
    - A LANE client for the *mkt* emulated LANs

In this scenario, once routing is enabled and network level addresses are assigned, Router 1 and Router 2 can route between the *man* and the *eng* emulated LANs, and Router 3 and Router 4 can route between the *man* and the *mkt* emulated LANs.

CHAPTER 21

Configuring LAN Emulation

This chapter describes how to configure LANE on the following platforms that are connected to an ATM switch or switch cloud:

● ATM Interface Processor (AIP) on the Cisco 7500 series routers

● ATM port adapter on the Cisco 7200 series and Cisco 7500 series routers

● Network Processor Module (NPM) on the Cisco 4500 and Cisco 4700 routers

---

NOTE   In Cisco IOS Release 11.3, all commands supported on the Cisco 7500 series routers are also supported on the Cisco 7000 series.

For a complete description of the LANE commands in this chapter, see Chapter 22, "LAN Emulation Commands." To locate documentation of other commands that appear in this chapter, you can search online at www.cisco.com.

LANE on ATM

LANE emulates an IEEE 802.3 Ethernet or IEEE 802.5 Token Ring LAN using ATM technology. LANE provides a service interface for network layer protocols that is identical to existing MAC layers. No changes are required to existing upper-layer protocols and applications. With LANE, Ethernet and Token Ring packets are encapsulated in the appropriate ATM cells and sent across the ATM network. When the packets reach the other side of the ATM network, they are de-encapsulated. LANE essentially bridges LAN traffic across ATM switches.

Benefits of LANE

ATM is a cell-switching and multiplexing technology designed to combine the benefits of circuit switching (constant transmission delay and guaranteed capacity) with those of packet switching (flexibility and efficiency for intermittent traffic).

LANE allows legacy Ethernet and Token Ring LAN users to take advantage of ATM's benefits without modifying end station hardware or software. ATM uses connection-oriented service with point-to-point signaling or multicast signaling between source and destination devices. However, LANs use connectionless service. Messages are broadcast to all devices on the network. With LANE, routers and switches emulate the connectionless service of a LAN for the end stations.

By using LANE, you can scale your networks to larger sizes while preserving your investment in LAN technology.

LANE Components

A single emulated LAN consists of the following entities: A LANE configuration server, a BUS, a LANE server, and LANE clients.

- **LANE configuration server**—A server that assigns individual clients to particular emulated LANs by directing them to the LANE server for the emulated LAN. The LECS maintains a database of LANE client and server ATM or MAC addresses and their emulated LANs. LECS can serve multiple emulated LANs.

- **BUS**—A multicast server that floods unknown destination traffic and forwards multicast and broadcast traffic to clients within an emulated LAN. One BUS exists per emulated LAN.

- **LANE server**—A server that provides a registration facility for clients to join the emulated LAN. There is one LES per emulated LAN. The LANE server handles LE ARP requests and maintains a list of LAN destination MAC addresses. For Token Ring LANE, the LANE server also maintains a list of route descriptors that is used to support source-route bridging over the emulated LAN. The route descriptors are used to determine the ATM address of the next hop in the RIF.

- **LANE client**—An entity in an endpoint, such as a router, that performs data forwarding, address resolution, and other control functions for a single endpoint in a single emulated LAN. The LEC provides standard LAN service to any higher layers that interface with it. A router can have multiple resident LANE clients, each connecting with different emulated LANs. The LANE client registers its MAC and ATM addresses with the LANE server.

Emulated LAN entities coexist on one or more Cisco routers. On Cisco routers, the LANE server and the BUS are combined into a single entity.

Other LANE components include ATM switches—any ATM switch that supports the ILMI and signaling. Multiple emulated LANs can coexist on a single ATM network.

Simple Server Redundancy

LANE relies on three servers: the LANE configuration server, the LANE server, and the BUS. If any one of these servers fails, the emulated LAN cannot fully function.

Cisco has developed a fault-tolerance mechanism known as *simple server redundancy* that eliminates these single points of failure. Although this scheme is proprietary, no new protocol additions have been made to the LANE subsystems.

Simple server redundancy uses multiple LANE configuration servers and multiple LANE BUSes. You can configure servers as backup servers, which become active if a master server fails. The priority levels for the servers determine which servers have precedence.

Refer to the section "Configuring the Fault-Tolerant Operation" for details and notes on the Simple Server Redundancy Protocol (SSRP).

Implementation Considerations

The following sections contain information relevant to implementation:

- Network support

- Hardware support

- Addressing

- Rules for assigning components to interfaces and subinterfaces

Network Support

In this release, Cisco supports the following networking features:

- Ethernet-emulated LANs

    - Routing from one emulated LAN to another via IP, IPX, or AppleTalk

    - Bridging between emulated LANs and between emulated LANs and other LANs

    - DECnet, Banyan VINES, and XNS routed protocols

- Token Ring emulated LANs

    - IP routing (fast switched) between emulated LANs and between a Token Ring emulated LAN and a legacy LAN

    - IPX routing between emulated LANs and between a Token Ring emulated LAN and a legacy LAN

    - Two-port and multiport source-route bridging (fast switched) between emulated LANs and between emulated LANs and a Token Ring

    - IP and IPX multiring

    - Source-route bridging (SRB), source-route translational bridging (SR/TLB), and source-route transparent bridging (SRT)

    - AppleTalk for (IOS) TR-LANE, which includes AppleTalk fast switched routing.

    - DECnet, Banyan VINES, and XNS protocols are not supported

    Cisco's implementation of LAN Emulation over 802.5 uses existing terminology and configuration options for Token Rings, including source-route bridging. Transparent bridging and Advanced Peer-to-Peer Networking (APPN) are not supported at this time.

- Hot Standby Router Protocol (HSRP)

Hardware Support

This release of LANE is supported on the following platforms:

- Cisco 4500-M, Cisco 4700-M

- Cisco 7200 series

- Cisco 7500 series

NOTE   In Cisco IOS Release 11.3, all commands supported on the Cisco 7500 series routers are also supported on the Cisco 7000 series routers equipped with RSP7000. Token Ring LAN emulation on Cisco 7000 series routers requires the RSP7000 upgrade. The RSP7000 upgrade requires a minimum of 24 MB DRAM and 8 MB Flash memory.

The router must contain an AIP, an ATM port adapter, or an NP-1A ATM NPM. These modules provide an ATM network interface for the routers. Network interfaces reside on modular interface processors, which provide a direct connection between the high-speed Cisco Extended Bus (CxBus) and the external networks. The maximum number of AIPs, ATM port adapters, or NPMs that the router supports depends on the bandwidth configured. The total bandwidth through all the AIPs, ATM port adapters, or NPMs in the system should be limited to 200 Mbps full duplex—two Transparent Asynchronous Transmitter/Receiver Interfaces (TAXIs), one Synchronous Optical Network (SONET) and one E3, or one SONET and one lightly used SONET.

This feature also requires one of the following switches:

- Cisco LightStream 1010 (recommended)

- Cisco LightStream 100

- Any ATM switch with UNI 3.0/3.1 and ILMI support for communicating the LECS address

TR-LANE requires Software Release 3.1(2) or later on the LightStream 100 switch and Cisco IOS Release 11.1(8) or later on the LightStream 1010.

For a complete description of the routers, switches, and interfaces, refer to your hardware documentation.

Addressing

On a LAN, packets are addressed by the MAC-layer address of the destination and source stations. To provide similar functionality for LANE, MAC-layer addressing must be supported. Every LANE client must have a MAC address. In addition, every LANE component (server, client, BUS, and configuration server) must have an ATM address that is different from that of all the other components.

All LANE clients on the same interface have the same, automatically assigned MAC address. That MAC address is also used as the end-system identifier (ESI) part of the ATM address, as explained in the next section. Although client MAC addresses are not unique, all ATM addresses are unique.

LANE ATM Addresses

A LANE ATM address has the same syntax as an NSAP, but it is not a network-level address. It consists of the following:

- A 13-byte prefix that includes the following fields defined by the ATM Forum:
    - AFI (Authority and Format Identifier) field (1 byte)
    - DCC (Data Country Code) or ICD (International Code Designator) field (2 bytes)
    - DFI field (Domain Specific Part Format Identifier) (1 byte)
    - Administrative Authority field (3 bytes)
    - Reserved field (2 bytes)
    - Routing Domain field (2 bytes)
    - Area field (2 bytes)
- A 6-byte ESI
- A 1-byte selector field

Cisco's Method of Automatically Assigning ATM Addresses

Cisco provides the following standard method of constructing and assigning ATM and MAC addresses for use in a LANE configuration server's database. A pool of MAC addresses is assigned to each ATM interface on the router. On Cisco 7200 series routers, Cisco 7500 series routers, Cisco 4500 routers, and Cisco 4700 routers, the pool contains eight MAC addresses. For constructing ATM addresses, the following assignments are made to the LANE components:

- The prefix fields are the same for all LANE components in the router; the prefix indicates the identity of the switch. The prefix value must be configured on the switch.
- The ESI field value assigned to every client on the interface is the first of the pool of MAC addresses assigned to the interface.
- The ESI field value assigned to every server on the interface is the second of the pool of MAC addresses.
- The ESI field value assigned to the BUS on the interface is the third of the pool of MAC addresses.
- The ESI field value assigned to the configuration server is the fourth of the pool of MAC addresses.

- The selector field value is set to the subinterface number of the LANE component—except for the LANE configuration server, which has a selector field value of 0.

  Because the LANE components are defined on different subinterfaces of an ATM interface, the value of the selector field in an ATM address is different for each component. The result is a unique ATM address for each LANE component, even within the same router. For more information about assigning components to subinterfaces, see the "Rules for Assigning Components to Interfaces and Subinterfaces" section later in this chapter.

For example, if the MAC addresses assigned to an interface are 0800.200C.1000 through 0800.200C.1007, the ESI part of the ATM addresses is assigned to LANE components as follows:

- Any client gets the ESI 0800.200c.1000.

- Any server gets the ESI 0800.200c.1001.

- The BUS gets the ESI 0800.200c.1002.

- The LANE configuration server gets the ESI 0800.200c.1003.

Refer to the "Multiple Token Ring Emulated LANs with Unrestricted Membership Example" and the "Multiple Token Ring Emulated LANs with Restricted Membership Example" sections for examples using MAC address values as ESI field values in ATM addresses, and for examples using subinterface numbers as selector field values in ATM addresses.

Using ATM Address Templates

ATM address templates can be used in many LANE commands that assign ATM addresses to LANE components (thus overriding automatically assigned ATM addresses) or that link client ATM addresses to emulated LANs. The use of templates can greatly simplify the use of these commands. The syntax of address templates, the use of address templates, and the use of wildcard characters within an address template for LANE are very similar to those for address templates of ISO CLNS.

NOTE   E.164-format ATM addresses do not support the use of LANE ATM address templates.

LANE ATM address templates can use two types of wildcard characters: an asterisk (*) to match any single character, and an ellipsis (...) to match any number of leading or trailing characters.

In LANE, a *prefix template* explicitly matches the prefix but uses wildcards for the ESI and selector fields. An *ESI template* explicitly matches the ESI field but uses wildcards for the prefix and selector.

Table 21-1 indicates how the values of unspecified digits are determined when an ATM address template is used.

Table 21-1   *Values of Unspecified Digits in ATM Address Templates*

Unspecified Digits In	Value Is
Prefix (first 13 bytes)	Obtained from ATM switch via ILMI
ESI (next 6 bytes)	Filled with the slot MAC address[1] plus
	• 0—LANE client
	• 1—LANE server
	• 2—LANE BUS
	• 3—Configuration server
Selector field (last 1 byte)	Subinterface number, in the range 0 through 255

1. The lowest of the pool of MAC addresses assigned to the ATM interface plus a value that indicates the LANE component. For the Cisco 7200 series routers, Cisco 7500 series routers, Cisco 4500 routers, and Cisco 4700 routers, the pool has eight MAC addresses.

Rules for Assigning Components to Interfaces and Subinterfaces

The following rules apply to assigning LANE components to the major ATM interface and its subinterfaces in a given router:

● The LANE configuration server always runs on the major interface. The assignment of any other component to the major interface is identical to assigning that component to the 0 subinterface.

● The server and the client of the *same* emulated LAN can be configured on the same subinterface in a router.

● Clients of two *different* emulated LANs cannot be configured on the same subinterface in a router.

● Servers of two *different* emulated LANs cannot be configured on the same subinterface in a router.

LANE Configuration Task List

Before you begin to configure LANE, you must decide whether you want to set up one or multiple emulated LANs. If you set up multiple emulated LANs, you must also decide where the servers and clients will be located, and whether to restrict the clients that can belong to each emulated LAN. Bridged emulated LANs are configured just like any other LAN, in terms of commands and outputs. Once you have made those basic decisions, you can proceed to configure LANE.

To configure LANE, complete the tasks in the following sections:

● Creating a LANE Plan and Worksheet

● Configuring the Prefix on the Switch

- Setting Up the Signaling and ILMI PVCs

- Displaying LANE Default Addresses

- Entering the Configuration Server's ATM Address on the Cisco Switch

- Setting Up the Configuration Server's Database

- Enabling the Configuration Server

- Setting Up LANE Servers and Clients

Once LANE is configured, you can configure MPOA. For MPOA to work with LANE, a LANE client must have an ELAN ID. To set up a LANE client for MPOA and give an ELAN ID, perform the tasks in the following section:

- Setting Up LANE Clients for MPOA

Although the sections listed here contain information about configuring SSRP fault tolerance, refer to the following section for detailed information about requirements and implementation considerations:

- Configuring the Fault-Tolerant Operation

When LANE is configured, you can monitor and maintain the components in the participating routers by completing the tasks in the following section:

- Monitoring and Maintaining the LANE Components

For configuration examples, see the "LANE Configuration Examples" section at the end of this chapter.

Creating a LANE Plan and Worksheet

It might help you to begin by drawing up a plan and a worksheet for your own LANE scenario, showing the following information and leaving space for noting the ATM address of each of the LANE components on each subinterface of each participating router:

- The router and interface where the LANE configuration server will be located

- The router, interface, and subinterface where the LANE server and BUS for each emulated LAN will be located. There can be multiple servers for each emulated LAN for fault-tolerant operation.

- The routers, interfaces, and subinterfaces where the clients for each emulated LAN will be located

- The name of the default emulated LAN (optional)

- The names of the emulated LANs that will have unrestricted membership

- The names of the emulated LANs that will have restricted membership

The last three items in this list are very important; they determine how you set up each emulated LAN in the configuration server's database.

Configuring the Prefix on the Switch

Before you configure LANE components on any Cisco 7200 series router, Cisco 7500 series router, Cisco 4500 router, or Cisco 4700 router, you must configure the Cisco ATM switch with the ATM address prefix to be used by all LANE components in the switch cloud. On the Cisco switch, the ATM address prefix is called the node ID. Prefixes must be 26 digits long. If you provide fewer than 26 digits, zeros are added to the right of the specified value to fill it to 26 digits.

To set the ATM address prefix on the Cisco LightStream 1010, use the following commands on the Cisco switch, beginning in global configuration mode:

Step	Command	Purpose
1	**atm address** {*atm-address*	*prefix...*}	Sets the local node ID (prefix of the ATM address).
2	**exit**	Exits global configuration mode.
3	**copy system:running-config nvram:startup-config**	Saves the configuration values permanently.

To set the ATM address prefix on the Cisco LightStream 100, use the following commands on the Cisco switch:

Step	Command	Purpose
1	**set local** *name ip-address mask prefix*	Sets the local node ID (prefix of the ATM address).
2	**save**	Saves the configuration values permanently.

On the Cisco switches, you can display the current prefix by using the **show network** command.

NOTE   If you do not save the configured value permanently, it will be lost when the switch is reset or powered off.

Setting Up the Signaling and ILMI PVCs

You must set up the signaling PVC and the PVC that will communicate with the ILMI on the major ATM interface of any router that participates in LANE.

Complete this task only once for a major interface. You do not need to repeat this task on the same interface even though you might configure LANE servers and clients on several of its subinterfaces.

To set up these PVCs, use the following commands, beginning in global configuration mode:

Step	Command	Purpose
1		Specifies the major ATM interface and enter interface configuration mode:
	interface atm *slot***/0**	On the AIP for Cisco 7500 series routers; On the ATM port adapter for Cisco 7200 series routers.
	interface atm *slot/port-adapter***/0**	On the ATM port adapter for Cisco 7500 series routers.
	interface atm *number*	On the NPM for Cisco 4500 and Cisco 4700 routers.
2	**atm pvc** *vcd vpi vci* **qsaal**	Sets up the signaling PVC that sets up and tears down SVCs; the *vpi* and *vci* values are usually set to 0 and 5, respectively.
3	**atm pvc** *vcd vpi vci* **ilmi**	Sets up a PVC to communicate with the ILMI; the *vpi* and *vci* values are usually set to 0 and 16, respectively.

Displaying LANE Default Addresses

You can display the LANE default addresses to make configuration easier. Complete this task for each router that participates in LANE. This command displays default addresses for all ATM interfaces present on the router. Write down the displayed addresses on your worksheet.

To display the default LANE addresses, use the following command in global configuration mode:

Command	Purpose
show lane default-atm-addresses	Displays the LANE default addresses.

Entering the Configuration Server's ATM Address on the Cisco Switch

You must enter the configuration server's ATM address into the Cisco LightStream 100 or Cisco LightStream 1010 ATM switch and save it permanently so that the value is not lost when the switch is reset or powered off.

You must specify the full 40-digit ATM address. Use the addresses on your worksheet that you obtained from the previous task.

If you are configuring SSRP, enter the multiple LANE configuration server addresses into the end ATM switches. The switches are used as central locations for the list of LANE configuration server addresses. LANE components connected to the switches obtain the global list of LANE configuration server addresses from the switches.

Depending on which type of switch you are using, perform one of the following tasks:

● Enter the ATM address(es) on the Cisco LightStream 1010 ATM switch

● Enter the ATM address(es) on the Cisco LightStream 100 ATM switch

Entering the ATM Address(es) on the Cisco LightStream 1010 ATM Switch

On the Cisco LightStream 1010 ATM switch, the configuration server address can be specified for a port or for the entire switch.

To enter the configuration server addresses on the Cisco LightStream 1010 ATM switch for the entire switch, use the following commands, beginning in global configuration mode:

Step	Command	Purpose
1	**atm lecs-address-default** *lecsaddress* [*sequence #*]	Specifies the LANE configuration server's ATM address for the entire switch. If you are configuring SSRP, include the ATM addresses of all the LANE configuration servers.
2	**exit**	Exits global configuration mode.
3	**copy system:running-config nvram:startup-config**	Saves the configuration value permanently.

To enter the configuration server addresses on the Cisco LightStream 1010 ATM switch per port, use the following commands beginning in interface configuration mode:

Step	Command	Purpose
1	**atm lecs-address** *lecsaddress* [*sequence #*]	Specifies the LANE configuration server's ATM address for a port. If you are configuring SSRP, include the ATM addresses of all the LANE configuration servers.
2	**Ctrl-Z**	Exits interface configuration mode.
3	**copy system:running-config nvram:startup-config**	Saves the configuration value permanently.

Entering the ATM Address(es) on the Cisco LightStream 100 ATM Switch

To enter the configuration server's ATM address into the Cisco LightStream 100 ATM switch and save it permanently, use the following commands in privileged EXEC mode:

Step	Command	Purpose
1	**set configserver** *index atm-address*	Specifies the LANE configuration server's ATM address. If you are configuring SSRP, repeat this command for each configuration server address. The *index* value determines the priority. The highest priority is 0. There can be a maximum of four LANE configuration servers.
2	**save**	Saves the configuration value permanently.

Setting Up the Configuration Server's Database

The configuration server's database contains information about each emulated LAN, including the ATM addresses of the LANE servers.

You can specify one default emulated LAN in the database. The LANE configuration server will assign any client that does not request a specific emulated LAN to the default emulated LAN.

Emulated LANs are either restricted or unrestricted. The configuration server will assign a client to an unrestricted emulated LAN if the client specifies that particular emulated LAN in its configuration. However, the configuration server will only assign a client to a restricted emulated LAN if the client is specified in the configuration server's database as belonging to that emulated LAN. The default emulated LAN must have unrestricted membership.

If you are configuring fault tolerance, you can have any number of servers per emulated LAN. Priority is determined by entry order; the first entry has the highest priority, unless you override it with the index option.

To set up the database, complete the tasks in the following sections as appropriate for your emulated LAN plan and scenario:

- Setting Up the Database for the Default Emulated LAN Only

- Setting Up the Database for Unrestricted-Membership Emulated LANs

- Setting Up the Database for Restricted-Membership LANs

Setting Up the Database for the Default Emulated LAN Only

When you configure a router as the configuration server for one default emulated LAN, you provide a name for the database, the ATM address of the LANE server for the emulated LAN, and a default name for the emulated LAN. In addition, you indicate that the configuration server's ATM address is to be computed automatically.

When you configure a database with only a default unrestricted emulated LAN, you do not have to specify where the LANE clients are located. That is, when you set up the configuration server's database for a single default emulated LAN, you do not have to provide any database entries that link the ATM addresses of any clients with the emulated LAN name. All of the clients will be assigned to the default emulated LAN.

To set up the configuration server for the default emulated LAN, use the following commands, beginning in global configuration mode:

Step	Command	Purpose
1	**lane database** *database-name*	Creates a named database for the LANE configuration server.

Step	Command	Purpose
2	**name** *elan-name* **server-atm-address** *atm-address* [**index** *number*]	In the configuration database, it binds the name of the emulated LAN to the ATM address of the LANE server.
		If you are configuring SSRP, repeat this step for each additional server for the same emulated LAN. The index determines the priority. The highest priority is 0.
3	**name** *elan-name* **local-seg-id** *segment-number*	If you are configuring a Token Ring emulated LAN, it assigns a segment number to the emulated Token Ring LAN in the configuration database.
4	**default-name** *elan-name*	In the configuration database, it provides a default name for the emulated LAN.
5	**exit**	Exits from database configuration mode and returns to global configuration mode.

In Step 2, enter the ATM address of the server for the specified emulated LAN, as noted in your worksheet and obtained in the "Displaying LANE Default Addresses" section.

You can have any number of servers per emulated LAN for fault tolerance. Priority is determined by entry order. The first entry has the highest priority unless you override it with the index option.

If you are setting up only a default emulated LAN, the *elan-name* value in Steps 2 and 3 is the same as the default emulated LAN name you provide in Step 4.

To set up fault-tolerant operation, see the "Configuring the Fault-Tolerant Operation" section later in this chapter.

Setting Up the Database for Unrestricted-Membership Emulated LANs

When you set up a database for unrestricted emulated LANs, you create database entries that link the name of each emulated LAN to the ATM address of its server.

However, you may choose not to specify where the LANE clients are located. That is, when you set up the configuration server's database, you do not have to provide any database entries that link the ATM addresses or MAC addresses of any clients with the emulated LAN name. The configuration server will assign the clients to the emulated LANs specified in the client's configurations.

To configure a router as the configuration server for multiple emulated LANs with unrestricted membership, use the following commands, beginning in global configuration mode:

Step	Command	Purpose
1	**lane database** *database-name*	Creates a named database for the LANE configuration server.
2	**name** *elan-name1* **server-atm-address** *atm-address* [**index** *number*]	In the configuration database, it binds the name of the first emulated LAN to the ATM address of the LANE server for that emulated LAN.
		If you are configuring SSRP, repeat this step with the same emulated LAN name but with different server ATM addresses for each additional server for the same emulated LAN. The index determines the priority. The highest priority is 0.
3	**name** *elan-name2* **server-atm-address** *atm-address* [**index** *number*]	In the configuration database, it binds the name of the second emulated LAN to the ATM address of the LANE server.
		If you are configuring SSRP, repeat this step with the same emulated LAN name but with different server ATM addresses for each additional server for the same emulated LAN. The index determines the priority. The highest priority is 0.
		Repeat this step, providing a different emulated LAN name and ATM address for each additional emulated LAN in this switch cloud.
4	**name** *elan-name1* **local-seg-id** *segment-number*	For a Token Ring emulated LAN, it assigns a segment number to the first emulated Token Ring LAN in the configuration database.
5	**name** *elan-name2* **local-seg-id** *segment-number*	For Token Ring emulated LANs, it assigns a segment number to the second emulated Token Ring LAN in the configuration database.
		Repeat this step, providing a different emulated LAN name and segment number for each additional source-route bridged emulated LAN in this switch cloud.
6	**default-name** *elan-name1*	(Optional) Specifies a default emulated LAN for LANE clients not explicitly bound to an emulated LAN.
7	**exit**	Exits database configuration mode and returns to global configuration mode.

In the preceding steps, enter the ATM address of the server for the specified emulated LAN, as noted in your worksheet and obtained in the "Displaying LANE Default Addresses" section.

To set up fault-tolerant operation, see the "Configuring the Fault-Tolerant Operation" section later in this chapter.

Setting Up the Database for Restricted-Membership LANs

When you set up the database for restricted-membership emulated LANs, you create database entries that link the name of each emulated LAN to the ATM address of its server.

However, you must also specify where the LANE clients are located. That is, for each restricted-membership emulated LAN, you provide a database entry that explicitly links the ATM address or MAC address of each client of that emulated LAN with the name of that emulated LAN.

The client database entries specify which clients are allowed to join the emulated LAN. When a client requests to join an emulated LAN, the configuration server consults its database and then assigns the client to the emulated LAN specified in the configuration server's database.

When clients for the same restricted-membership emulated LAN are located in multiple routers, each client's ATM address or MAC address must be linked explicitly with the name of the emulated LAN. As a result, you must configure as many client entries (at Steps 6 and 7, in the following procedure) as you have clients for emulated LANs in all the routers. Each client will have a different ATM address in the database entries.

To set up the configuration server for emulated LANs with restricted membership, use the following commands, beginning in global configuration mode:

Step	Command	Purpose
1	**lane database** *database-name*	Creates a named database for the LANE configuration server.
2	**name** *elan-name1* **server-atm-address** *atm-address* **restricted** [**index** *number*]	In the configuration database, it binds the name of the first emulated LAN to the ATM address of the LANE server for that emulated LAN.
		If you are configuring SSRP, repeat this step with the same emulated LAN name but with different server ATM addresses for each additional server for the same emulated LAN. The index determines the priority. The highest priority is 0.
3	**name** *elan-name2* **server-atm-address** *atm-address* **restricted** [**index** *number*]	In the configuration database, it binds the name of the second emulated LAN to the ATM address of the LANE server.
		If you are configuring SSRP, repeat this step with the same emulated LAN name but with different server ATM addresses for each additional server for the same emulated LAN. The index determines the priority. The highest priority is 0.
		Repeat this step, providing a different name and a different ATM address, for each additional emulated LAN.
4	**name** *elan-name1* **local-seg-id** *segment-number*	For a Token Ring emulated LAN, it assigns a segment number to the first emulated Token Ring LAN in the configuration database.

5	**name** *elan-name2* **local-seg-id** *segment-number*	If you are configuring Token Ring emulated LANs, it assigns a segment number to the second emulated Token Ring LAN in the configuration database.
		Repeat this step, providing a different emulated LAN name and segment number for each additional source-route bridged emulated LAN in this switch cloud.
6	**client-atm-address** *atm-address-template* **name** *elan-name1*	Adds a database entry associating a specific client's ATM address with the first restricted-membership emulated LAN.
		Repeat this step for each of the clients of the first restricted-membership emulated LAN.
7	**client-atm-address** *atm-address-template* **name** *elan-name2*	Adds a database entry associating a specific client's ATM address with the second restricted-membership emulated LAN.
		Repeat this step for each of the clients of the second restricted-membership emulated LAN.
		Repeat this step, providing a different name and a different list of client ATM address, for each additional emulated LAN.
8	**exit**	Exits database configuration mode and returns to global configuration mode.

To set up fault-tolerant operation, see the "Configuring the Fault-Tolerant Operation" section later in this chapter.

Enabling the Configuration Server

Once you have created the database, you can enable the configuration server on the selected ATM interface and router by using the following commands, beginning in global configuration mode:

Step	Command	Purpose
1		If you are not currently configuring the interface, it specifies the major ATM interface where the configuration server is located:
	interface atm *slot*/**0**[.*subinterface-number*]	On the AIP for Cisco 7500 series routers and on the ATM port adapter for Cisco 7200 series routers.
		On the ATM port adapter for Cisco 7500 series routers.
		On the NPM for Cisco 4500 and Cisco 4700 routers.
	interface atm *slot/port-adapter/***0**[.*subinterface-number*]	
	interface atm *number*[.*subinterface-number*]	

Step	Command	Purpose
2	**lane config database** *database-name*	Links the configuration server's database name to the specified major interface, and enables the configuration server.
3		Specifies how the LECS's ATM address will be computed. You may opt to choose one of the following scenarios:
	lane config auto-config-atm-address	The LECS will participate in SSRP and the address is computed by the automatic method.
	lane config auto-config-atm-address **lane config fixed-config-atm-address**	The LECS will participate in SSRP, and the address is computed by thc automatic method. If the LECS is the master, the fixed address is also used.
		The LECS will not participate in SSRP, the LECS is the master, and only the well-known address is used.
	lane config fixed-config-atm-address	The LECS will participate in SSRP and the address is computed using an explicit, 20-byte ATM address.
	lane config config-atm-address *atm-address-template*	
4	**exit**	Exits interface configuration mode.
5	**Ctrl-Z**	Returns to EXEC mode.
6	**copy system:running-config nvram:startup-config**	Saves the configuration.

Setting Up LANE Servers and Clients

For each router that will participate in LANE, set up the necessary servers and clients for each emulated LAN; then display and record the server and client ATM addresses. Be sure to keep track of the router interface where the LANE configuration server will eventually be located.

You can set up servers for more than one emulated LAN on different subinterfaces or on the same interface of a router, or you can place the servers on different routers.

When you set up a server and BUS on a router, you can combine them with a client on the same subinterface, a client on a different subinterface, or no client at all on the router.

Where you put the clients is important because any router with clients for multiple emulated LANs can route frames between those emulated LANs.

Depending on where your clients and servers are located, perform one of the following tasks for each LANE subinterface.

- Set up the server, a BUS, and a client on a subinterface

- Set up only a client on a subinterface

Setting Up the Server, a BUS, and a Client on a Subinterface

To set up the server, a BUS, and (optionally) clients for an emulated LAN, use the following commands, beginning in global configuration mode:

Step	Command	Purpose
1		Specifies the subinterface for the emulated LAN on this router:
	interface atm *slot*/**0**.*subinterface-number*	On the AIP for Cisco 7500 series routers, and on the ATM port adapter for Cisco 7200 series routers.
		On the ATM port adapter for Cisco 7500 series routers.
	interface atm *slot*/*port-adapter*/ **0**.*subinterface-number*	On the NPM for Cisco 4500 and Cisco 4700 routers.
	interface atm *number.subinterface-number*	
2	**lane server-bus** {**ethernet**	**tokenring**} *elan-name*	Enables a LANE server and a LANE BUS for the emulated LAN.
3	**lane client** {**ethernet**	**tokenring**} [*elan-name*] [**elan-id** *id*]	(Optional) Enables a LANE client for the emulated LAN.
		To participate in MPOA, configure the LANE server and a LANE BUS for the emulated LAN with the ELAN ID.
4	**ip** *address mask*[1]	Provides a protocol address for the client.
5	**Ctrl-Z**	Returns to EXEC mode.
6	**copy system:running-config nvram:startup-config**	Saves the configuration.

1. The command or commands depend on the routing protocol used.

If the emulated LAN in Step 3 is intended to have *restricted membership*, consider carefully whether you want to specify its name here. You will specify the name in the LANE configuration server's database when it is set up. However, if you link the client to an emulated LAN in this step, and through some mistake it does not match the database entry linking the client to an emulated LAN, this client will not be allowed to join this emulated LAN or any other.

If you do decide to include the name of the emulated LAN linked to the client in Step 3 and later want to associate that client with a different emulated LAN, make the change in the configuration server's database before you make the change for the client on this subinterface.

Each emulated LAN is a separate subnetwork. In Step 4 make sure that the clients of the same emulated LAN are assigned protocol addresses on the same subnetwork and that clients of different emulated LANs are assigned protocol addresses on different subnetworks.

Setting Up Only a Client on a Subinterface

On any given router, you can set up one client for one emulated LAN or multiple clients for multiple emulated LANs. You can set up a client for a given emulated LAN on any routers you choose to participate in that emulated LAN. Any router with clients for multiple emulated LANs can route packets between those emulated LANs.

You must first set up the signaling and ILMI PVCs on the major ATM interface, as described earlier in the "Setting Up the Signaling and ILMI PVCs" section, before you set up the client.

To set up only a client for an emulated LAN, use the following commands, beginning in interface configuration mode:

Step	Command	Purpose
1		Specifies the subinterface for the emulated LAN on this router:
	interface atm *slot/***0.***subinterface-number*	On the AIP for Cisco 7500 series routers, and on the ATM port adapter for Cisco 7200 series routers.
	interface atm *slot/port-adapter/***0.***subinterface-number*	On the ATM port adapter for Cisco 7500 series routers.
	interface atm *number.subinterface-number*	On the NPM for Cisco 4500 and Cisco 4700 routers.
2	**ip** *address mask*[1]	Provides a protocol address for the client on this subinterface.
3	**lane client** {**ethernet** I **tokenring**} [*elan-name*]	Enables a LANE client for the emulated LAN.
4	**Ctrl-Z**	Returns to EXEC mode.
5	**copy system:running-config nvram:startup-config**	Saves the configuration.

1. The command or commands depend on the routing protocol used.

Each emulated LAN is a separate subnetwork. In Step 2, make sure that the clients of the same emulated LAN are assigned protocol addresses on the same subnetwork, and that clients of different emulated LANs are assigned protocol addresses on different subnetworks.

Setting Up LANE Clients for MPOA

For MPOA to work properly, a LANE client must have an ELAN ID for all ELANs represented by the LANE client. To configure an ELAN ID, use one of the following commands in LANE database configuration mode or in interface configuration mode when starting up the LES for that emulated LAN:

Command	Purpose
name *elan-name* **elan-id** *id*	Configures the ELAN ID in the LECS database to participate in MPOA.
lane server-bus {**ethernet** | **tokenring**} *elan-name* [**elan-id** *id*]	Configures the LANE server and a LANE broadcast-and-unknown server for the emulated LAN.
	To participate in MPOA, configure the LANE server and a LANE broadcast-and-unknown server for the emulated LAN with the ELAN ID.

CAUTION   If an ELAN ID is supplied by both commands, make sure that the ELAN ID matches in both.

Configuring the Fault-Tolerant Operation

The LANE simple server redundancy feature creates fault tolerance using standard LANE protocols and mechanisms. If a failure occurs on the LANE configuration server or on the LANE server/broadcast-and-unknown server, the emulated LAN can continue to operate using the services of a backup LANE server. This protocol is called SSRP.

This section describes how to configure simple server redundancy for fault tolerance on an emulated LAN.

NOTE   This server redundancy does not overcome other points of failure beyond the router ports: Additional redundancy on the LAN side or in the ATM switch cloud are not a part of the LANE simple server redundancy feature.

Simple Server Redundancy Requirements

For simple LANE service replication or fault tolerance to work, the ATM switch must support multiple LANE server addresses. This mechanism is specified in the LANE standard. The LANE servers establish and maintain a standard control circuit that enables the server redundancy to operate.

LANE simple server redundancy is supported on Cisco IOS Release 11.2 and later software. Older LANE configuration files continue to work with this new software.

This redundancy feature works only with Cisco LANE configuration servers and LANE server/BUS combinations. Third-party LANE Clients can be used with the SSRP, but third-party configuration servers, LANE servers, and BUSes do not support SSRP.

For server redundancy to work correctly:

● All the ATM switches must have identical lists of the global LANE configuration server addresses, in the identical priority order.

● The operating LANE configuration servers must use exactly the same configuration database. Load the configuration table data using the **copy** {**rcp** | **tftp**} **system:running-config** command. This method minimizes errors and enables the database to be maintained centrally in one place.

The LANE protocol does not specify where any of the emulated LAN server entities should be located, but for the purpose of reliability and performance, Cisco implements these server components on its routers.

Redundant Configuration Servers

To enable redundant LANE configuration servers, enter the multiple LANE configuration server addresses into the end ATM switches. LANE components can obtain the list of LANE configuration server addresses from the ATM switches through the ILMI.

Refer to the "Entering the Configuration Server's ATM Address on the Cisco Switch" section for more details.

Redundant Servers and BUSes

The LANE configuration server turns on server/BUS redundancy by adjusting its database to accommodate multiple server ATM addresses for a particular emulated LAN. The additional servers serve as backup servers for that emulated LAN.

To activate the feature, you add an entry for the hierarchical list of servers that will support the given emulated LAN. All database modifications for the emulated LAN must be identical on all LANE configuration servers.

Refer to the "Setting Up the Configuration Server's Database" section for more details.

Implementation Considerations

● The LightStream 1010 can handle up to 16 LANE configuration server addresses. The LightStream 100 allows a maximum of four LANE configuration server addresses.

● There is no limit on the number of LANE servers that can be defined per emulated LAN.

- When a LANE configuration server switchover occurs, no previously joined clients are affected.

- When a LANE server/BUS switches over, momentary loss of clients occurs until they are all transferred to the new LANE server/BUS.

- LANE configuration servers come up as masters until a higher-level LANE configuration server tells them otherwise. This is automatic and cannot be changed.

- If a higher-priority LANE server comes online, it bumps the current LANE server off on the same emulated LAN. Therefore, there may be some flapping of clients from one LANE server to another after a powerup, depending on the order of the LANE servers coming up. Flapping should settle after the *last* highest-priority LANE server comes up.

- If none of the specified LANE servers are up or connected to the master LANE configuration server and more than one LANE server is defined for an emulated LAN, a configuration request for that specific emulated LAN is rejected by the LANE configuration server.

- Changes made to the list of LANE configuration server addresses on ATM switches may take up to a minute to propagate through the network. Changes made to the configuration database regarding LANE server addresses take effect almost immediately.

- If none of the designated LANE configuration servers are operational or reachable, the ATM Forum-defined well-known LANE configuration server address is used.

- You can override the LANE configuration server address on any subinterface, by using the following commands:

  — **lane auto-config-atm-address**

  — **lane fixed-config-atm-address**

  — **lane config-atm-address**

CAUTION   When an override like this is performed, fault-tolerant operation cannot be guaranteed. To avoid affecting the fault-tolerant operation, do not override any LANE configuration server, LANE server, or BUS addresses.

- If an underlying ATM network failure occurs, there may be multiple master LANE configuration servers and multiple active LANE servers for the same emulated LAN. This situation creates a "partitioned" network. The clients continue to operate normally, but transmission between different partitions of the network is not possible. When the network break is repaired, the system recovers.

- When the LECS is already up and running, and you use the **lane config fixed-config-atm-address** command to configure the well-known LECS address, be aware of the following scenarios:

— If you configure the LECS with only the well-known address, the LECS will not participate in the SSRP; it will act as a "standalone" master, and only listen on the well-known LECS address. This scenario is ideal if you want a standalone LECS that does not participate in SSRP, and you would like to listen to only the well-known address.

— If only the well-known address is already assigned, and you assign at least one other address to the LECS, (additional addresses are assigned using the **lane config auto-config-atm-address** command and/or the **lane config config-atm-address command**) the LECS will participate in the SSRP and act as the master or slave based on the normal SSRP rules. This scenario is ideal if you would like the LECS to participate in SSRP, and you would like to make the master LECS listen on the well-known address.

— If the LECS is participating in SSRP, has more than one address (one of which is the well-known address), and all the addresses but the well-known address is removed, the LECS will declare itself the master and stop participating in SSRP completely.

— If the LECS is operating as an SSRP slave, and it has the well-known address configured, it will not listen on the well-known address unless it becomes the master.

— If you want the LECS to assume the well-known address only when it becomes the master, configure the LECS with the well-known address and at least one other address.

SSRP Changes to Reduce Network Flap

SSRP was originally designed so that when a higher LES came on line, all the LECs in that emulated LAN flipped over to the higher LES. This caused unnecessary disruptions in large networks. Now SSRP is designed to eliminate unnecessary flapping. If the current LES is healthy, the flapping can be eliminated by changing the SSRP behavior so that the emulated LAN does not flip over to another LES. Obviously, if the currently active LES goes down, all the LECs will then be switched over to the first available highest LES in the list. This is now the default behavior.

If emulated LANs are now configured in the new way, a LECS switchover may or may not cause a network flap depending on how quickly each LES now reconnects to the new master LECS. If the old active LES connects first, the flap will not occur. However, if another LES connects first (since now the criteria is that the first connected LES is assumed the master LES, rather than the highest-ranking one), then the network will still flap.

For customers who would specifically like to maintain the old SSRP behavior, they can use the new LECS **name** *elan-name* **preempt** command. This command will force the old behavior to be maintained. This feature can be enabled/disabled on a per individual emulated LAN basis from the LECS database. In the older scheme (preempt), the LES switchover caused network flap.

To enable network flap and set the emulated LAN preempt for a LANE server, use the following command in LANE database configuration mode:

Command	Purpose
name *elan-name* **preempt**	Sets the emulated LAN LES preemption.

Monitoring and Maintaining the LANE Components

After configuring LANE components on an interface or any of its subinterfaces, on a specified subinterface, or on an emulated LAN, you can display their status. To show LANE information, use the following commands in EXEC mode:

Command	Purpose
	Displays the global and per-virtual channel connection LANE information for all the LANE components and emulated LANs configured on an interface or any of its subinterfaces:
show lane [**interface atm** *slot*/**0**[*.subinterface-number*] I **name** *elan-name*] [**brief**]	• On the AIP for Cisco 7500 series routers, and on the ATM port adapter for Cisco 7200 series routers.
show lane [**interface atm** *slot*/*port-adapter*/**0**[*.subinterface-number*] I **name** *elan-name*] [**brief**]	• On the ATM port adapter for Cisco 7500 series routers.
show lane [**interface atm** *number*[*.subinterface-number*] I **name** *elan-name*] [**brief**]	• On the NPM for Cisco 4500 and Cisco 4700 routers.
	Displays the global and per-VCC LANE information for the broadcast-and-unknown server configured on any subinterface or emulated LAN:
show lane bus [**interface atm** *slot*/**0**[*.subinterface-number*] I **name** *elan-name*] [**brief**]	• On the AIP for Cisco 7500 series routers, and on the ATM port adapter for Cisco 7200 series routers.
show lane bus [**interface atm** *slot*/*port-adapter*/ **0** [*.subinterface-number*] I **name** *elan-name*] [**brief**]	• On the ATM port adapter for Cisco 7500 series routers.
show lane bus [**interface atm** *number*[*.subinterface-number*] I **name** *elan-name*] [**brief**]	• On the NPM for Cisco 4500 and Cisco 4700 routers.

Command	Purpose
	Displays the global and per-VCC LANE information for all LANE clients configured on any subinterface or emulated LAN:
show lane client [**interface atm** *slot***/0**[*.subinterface-number*] | **name** *elan-name*] [**brief**]	• On the AIP for Cisco 7500 series routers, and on the ATM port adapter for Cisco 7200 series routers.
show lane client [**interface atm** *slot/port-adapter***/0**[*.subinterface-number*] | **name** *elan-name*] [**brief**]	• On the ATM port adapter for Cisco 7500 series routers.
show lane client [**interface atm** *number*[*.subinterface-number*] | **name** *elan-name*] [**brief**]	• On the NPM for Cisco 4500 and Cisco 4700 routers.
	Displays the global and per-VCC LANE information for the configuration server configured on any interface:
show lane config [**interface atm** *slot***/0**]	• On the AIP for Cisco 7500 series routers, and on the ATM port adapter for Cisco 7200 series routers.
show lane config [**interface atm** *slot/port-adapter***/0**]	• On the ATM port adapter for Cisco 7500 series routers.
show lane config [**interface atm** *number*]	• On the NPM for Cisco 4500 and Cisco 4700 routers.
show lane database [*database-name*]	Displays the LANE configuration server's database.
	Displays the automatically assigned ATM address of each LANE component in a router or on a specified interfacc or subinterface:
show lane default-atm-addresses [**interface atm** *slot***/0.***subinterface-number*]	• On the AIP for Cisco 7500 series routers, and on the ATM port adapter for Cisco 7200 series routers.
show lane default-atm-addresses [**interface atm** *slot/port-adapter/***0.***subinterface-number*]	• On the ATM port adapter for Cisco 7500 series routers.
show lane default-atm-addresses [**interface atm** *number.subinterface-number*]	• On the NPM for Cisco 4500 and Cisco 4700 routers.

Command	Purpose
	Displays the LANE ARP table of the LANE client configured on the specified subinterface or emulated LAN:
show lane le-arp [**interface atm** *slot*/**0**[.*subinterface-number*]	**name** *elan-name*]	• On the AIP for Cisco 7500 series routers, and on the ATM port adapter for Cisco 7200 series routers.
show lane le-arp [**interface atm** *slot*/*port-adapter*/**0**[.*subinterface-number*]	**name** *elan-name*]	• On the ATM port adapter for Cisco 7500 series routers.
show lane le-arp [**interface atm** *number*[.*subinterface-number*]	**name** *elan-name*]	• On the NPM for Cisco 4500 and Cisco 4700 routers.
	Displays the global and per-VCC LANE information for the LANE server configured on a specified subinterface or emulated LAN:
show lane server [**interface atm** *slot*/**0**[.*subinterface-number*]	**name** *elan-name*] [**brief**]	• On the AIP for Cisco 7500 series routers, and on the ATM port adapter for Cisco 7200 series routers.
show lane server [**interface atm** *slot*/*port-adapter*/**0**[.*subinterface-number*]	**name** *elan-name*] [**brief**]	• On the ATM port adapter for Cisco 7500 series routers.
show lane server [**interface atm** *number*[.*subinterface-number*]	**name** *elan-name*] [**brief**]	• On the NPM for Cisco 4500 and Cisco 4700 routers.

LANE Configuration Examples

The examples in the following sections illustrate how to configure LANE for the following cases:

- Default Configuration for a Single Ethernet Emulated LAN Example

- Default Configuration for a Single Ethernet Emulated LAN with a Backup LANE Configuration Server and LANE Server Example

- Multiple Token Ring Emulated LANs with Unrestricted Membership Example

- Multiple Token Ring Emulated LANs with Restricted Membership Example

- TR-LANE with Two-Port Source-Route Bridging Example

- TR-LANE with Multiport Source-Route Bridging Example

- Routing Between Token Ring and Ethernet Emulated LANs Example

All examples use the automatic ATM address assignment method described in the "Cisco's Method of Automatically Assigning ATM Addresses" section earlier in this chapter. These examples show the LANE configurations, not the process of determining the ATM addresses and entering them.

Default Configuration for a Single Ethernet Emulated LAN Example

The following example configures four Cisco 7500 series routers for one Ethernet emulated LAN. Router 1 contains the configuration server, the server, the BUS, and a client. The remaining routers each contain a client for the emulated LAN. This example accepts all default settings that are provided. For example, it does not explicitly set ATM addresses for the different LANE components that are colocated on the router. Membership in this LAN is not restricted.

The following is an example of Router 1 configuration:

```
lane database example1
 name eng server-atm-address 39.000001415555121101020304.0800.200c.1001.01
 default-name eng
interface atm 1/0
 atm pvc 1 0 5 qsaal
 atm pvc 2 0 16 ilmi
 lane config auto-config-atm-address
 lane config database example1
interface atm 1/0.1
 ip address 172.16.0.1 255.255.255.0
 lane server-bus ethernet eng
 lane client ethernet
```

The following is an example of Router 2 configuration:

```
interface atm 1/0
 atm pvc 1 0 5 qsaal
 atm pvc 2 0 16 ilmi
interface atm 1/0.1
 ip address 172.16.0.3 255.255.255.0
 lane client ethernet
```

The following is an example of Router 3 configuration:

```
interface atm 2/0
 atm pvc 1 0 5 qsaal
 atm pvc 2 0 16 ilmi
interface atm 2/0.1
 ip address 172.16.0.4 255.255.255.0
 lane client ethernet
```

The following is an example of Router 4 configuration:

```
interface atm 1/0
 atm pvc 1 0 5 qsaal
 atm pvc 2 0 16 ilmi
interface atm 1/0.3
 ip address 172.16.0.5 255.255.255.0
 lane client ethernet
```

Default Configuration for a Single Ethernet Emulated LAN with a Backup LANE Configuration Server and LANE Server Example

This example configures four Cisco 7500 series routers for one emulated LAN with fault tolerance. Router 1 contains the configuration server, the server, the BUS, and a client. Router 2 contains the backup LANE configuration server and the backup LANE server for this emulated LAN and another client. Routers 3 and 4 contain clients only. This example accepts all default settings that are provided. For example, it does not explicitly set ATM addresses for the various LANE components colocated on the router. Membership in this LAN is not restricted.

The following is an example of Router 1 configuration:

```
lane database example1
 name eng server-atm-address 39.000001415555121101020304.0800.200c.1001.01
 name eng server-atm-address 39.000001415555121101020304.0612.200c 2001.01
 default-name eng
interface atm 1/0
 atm pvc 1 0 5 qsaal
 atm pvc 2 0 16 ilmi
 lane config auto-config-atm-address
 lane config database example1
interface atm 1/0.1
 ip address 172.16.0.1 255.255.255.0
 lane server-bus ethernet eng
 lane client ethernet
```

The following is an example of Router 2 configuration:

```
lane database example1_backup
 name eng server-atm-address 39.000001415555121101020304.0800.200c.1001.01
 name eng server-atm-address 39.000001415555121101020304.0612.200c 2001.01 (backup LES)
 default-name eng
interface atm 1/0
 atm pvc 1 0 5 qsaal
 atm pvc 2 0 16 ilmi
 lane config auto-config-atm-address
 lane config database example1_backup
interface atm 1/0.1
 ip address 172.16.0.3 255.255.255.0
 lane server-bus ethernet eng
 lane client ethernet
```

The following is an example of Router 3 configuration:

```
interface atm 2/0
 atm pvc 1 0 5 qsaal
 atm pvc 2 0 16 ilmi
interface atm 2/0.1
 ip address 172.16.0.4 255.255.255.0
 lane client ethernet
```

The following is an example of Router 4 configuration:

```
interface atm 1/0
 atm pvc 1 0 5 qsaal
 atm pvc 2 0 16 ilmi
interface atm 1/0.3
 ip address 172.16.0.5 255.255.255.0
 lane client ethernet
```

Multiple Token Ring Emulated LANs with Unrestricted Membership Example

The following example configures four Cisco 7500 series routers for three emulated LANS for engineering, manufacturing, and marketing, as illustrated in Figure 21-1. This example does not restrict membership in the emulated LANs.

Figure 21-1 *Multiple Emulated LANs*

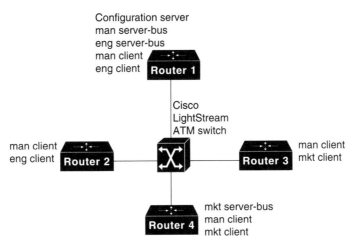

In this example, Router 1 has the following LANE components:

● The LANE configuration server (there is one configuration server for this group of emulated LANs)

● The LANE server and BUS for the emulated LAN for manufacturing (*man*)

● The LANE server and BUS for the emulated LAN for engineering (*eng*)

● A LANE client for the emulated LAN for manufacturing (*man*)

● A LANE client for the emulated LAN for engineering (*eng*)

Router 2 has the following LANE components:

● A LANE client for the emulated LAN for manufacturing (*man*)

● A LANE client for the emulated LAN for engineering (*eng*)

Router 3 has the following LANE components:

- A LANE client for the emulated LAN for manufacturing (*man*)

- A LANE client for the emulated LAN for marketing (*mkt*)

Router 4 has the following LANE components:

- The LANE server and BUS for the emulated LAN for marketing (*mkt*)

- A LANE client for the emulated LAN for manufacturing (*man*)

- A LANE client for the emulated LAN for marketing (*mkt*)

For the purposes of this example, the four routers are assigned ATM address prefixes and ESIs as shown in Table 21-2 (the ESI part of the ATM address is derived from the first MAC address of the AIP shown in the example).

Table 21-2 *ATM Prefixes for TR-LANE Example*

Router	ATM Address Prefix	ESI Base
Router 1	39.000001415555121101020304	0800.200c.1000
Router 2	39.000001415555121101020304	0800.200c.2000
Router 3	39.000001415555121101020304	0800.200c.3000
Router 4	39.000001415555121101020304	0800.200c.4000

Router 1

Router 1 has the configuration server and its database, the server and BUS for the manufacturing emulated LAN, the server and BUS for the engineering emulated LAN, a client for manufacturing, and a client for engineering. Router 1 is configured as shown in this example:

```
!The following lines name and configure the configuration server's database.
lane database example2
 name eng server-atm-address 39.000001415555121101020304.0800.200c.1001.02
 name eng local-seg-id 1000
 name man server-atm-address 39.000001415555121101020304.0800.200c.1001.01
 name man local-seg-id 2000
 name mkt server-atm-address 39.000001415555121101020304.0800.200c.4001.01
 name mkt local-seg-id 3000
 default-name man
!
! The following lines bring up the configuration server and associate
! it with a database name.
interface atm 1/0
 atm pvc 1 0 5 qsaal
 atm pvc 2 0 16 ilmi
 lane config auto-config-atm-address
 lane config database example2
!
```

```
! The following lines configure the "man" server, broadcast-and-unknown server,
! and the client on atm subinterface 1/0.1. The client is assigned to the default
! emulated lan.
interface atm 1/0.1
 ip address 172.16.0.1 255.255.255.0
 lane server-bus tokenring man
 lane client tokenring man
!
! The following lines configure the "eng" server, broadcast-and-unknown server,
! and the client on atm subinterface 1/0.2. The client is assigned to the
! engineering emulated lan. Each emulated LAN is a different subnetwork, so the "eng"
! client has an IP address on a different subnetwork than the "man" client.
interface atm 1/0.2
 ip address 172.16.1.1 255.255.255.0
 lane server-bus tokenring eng
 lane client tokenring eng
```

Router 2

Router 2 is configured for a client of the manufacturing emulated LAN and a client of the engineering emulated LAN. Because the default emulated LAN name is *man*, the first client is linked to that emulated LAN name by default. Router 2 is configured as shown here:

```
interface atm 1/0
 atm pvc 1 0 5 qsaal
 atm pvc 2 0 16 ilmi
interface atm 1/0.1
 ip address 172.16.0.2 255.255.255.0
 lane client tokenring
interface atm 1/0.2
 ip address 172.16.1.2 255.255.255.0
 lane client tokenring eng
```

Router 3

Router 3 is configured for a client of the manufacturing emulated LAN and a client of the marketing emulated LAN. Because the default emulated LAN name is *man*, the first client is linked to that emulated LAN name by default. Router 3 is configured as shown here:

```
interface atm 2/0
 atm pvc 1 0 5 qsaal
 atm pvc 2 0 16 ilmi
interface atm 2/0.1
 ip address 172.16.0.3 255.255.255.0
 lane client tokenring
interface atm 2/0.2
 ip address 172.16.2.3 255.255.255.0
 lane client tokenring mkt
```

Router 4

Router 4 has the server and BUS for the marketing emulated LAN, a client for marketing, and a client for manufacturing. Because the default emulated LAN name is *man*, the second client is linked to that emulated LAN name by default. Router 4 is configured as shown here:

```
interface atm 3/0
 atm pvc 1 0 5 qsaal
 atm pvc 2 0 16 ilmi
interface atm 3/0.1
 ip address 172.16.2.4 255.255.255.0
 lane server-bus tokenring mkt
 lane client tokenring mkt
interface atm 3/0.2
 ip address 172.16.0.4 255.255.255.0
 lane client tokenring
```

Multiple Token Ring Emulated LANs with Restricted Membership Example

The following example, illustrated in Figure 21-2, configures a Cisco 7500 series router for three emulated LANS for engineering, manufacturing, and marketing.

The same components are assigned to the four routers as in the previous example. The ATM address prefixes and MAC addresses are also the same as in the previous example.

However, this example restricts membership for the engineering and marketing emulated LANs. The LANE configuration server's database has explicit entries binding the ATM addresses of LANE clients to specified, named emulated LANs. In such cases, the client requests information from the configuration server about which emulated LAN it should join; the configuration server checks its database and replies to the client. Because the manufacturing emulated LAN is unrestricted, any client not in the LANE configuration server's database is allowed to join it.

Figure 21-2   *Multiple Emulated LANs with Restricted Membership*

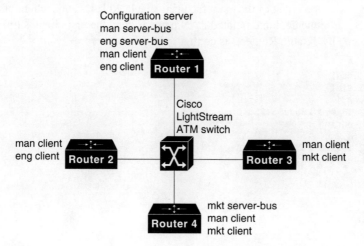

Router 1

Router 1 has the configuration server and its database, the server and BUS for the manufacturing emulated LAN, the server and broadcast-and-unknown server for the engineering emulated LAN, a client for manufacturing, and a client for engineering. It also has explicit database entries binding the ATM addresses of LANE clients to specified, named emulated LANs. Router 1 is configured as shown here:

```
! The following lines name and configure the configuration server's database.
lane database example3
 name eng server-atm-address 39.000001415555121101020304.0800.200c.1001.02 restricted
 name eng local-seg-id 1000
 name man server-atm-address 39.000001415555121101020304.0800.200c.1001.01
 name man local-seg-id 2000
 name mkt server-atm-address 39.000001415555121101020304.0800.200c.4001.01 restricted
 name mkt local-seg-id 3000
!
 ! The following lines add database entries binding specified client ATM
 ! addresses to emulated LANs. In each case, the Selector byte corresponds
 ! to the subinterface number on the specified router.
 ! The next command binds the client on Router 1's subinterface 2 to the eng ELAN.
 client-atm-address 39.0000014155551211.0800.200c.1000.02 name eng
 ! The next command binds the client on Router 2's subinterface 2 to the eng ELAN.
 client-atm-address 39.0000014155551211.0800.200c.2000.02 name eng
 ! The next command binds the client on Router 3's subinterface 2 to the mkt ELAN.
 client-atm-address 39.0000014155551211.0800.200c.3000.02 name mkt
 ! The next command binds the client on Router 4's subinterface 1 to the mkt ELAN.
 client-atm-address 39.0000014155551211.0800.200c.4000.01 name mkt
 default-name man
!
! The following lines bring up the configuration server and associate
! it with a database name.
interface atm 1/0
 atm pvc 1 0 5 qsaal
 atm pvc 2 0 16 ilmi
 lane config auto-config-atm-address
 lane config database example3
!
! The following lines configure the "man" server/broadcast-and-unknown server,
! and the client on atm subinterface 1/0.1. The client is assigned to the default
! emulated lan.
interface atm 1/0.1
 ip address 172.16.0.1 255.255.255.0
 lane server-bus tokenring man
 lane client tokenring
!
! The following lines configure the "eng" server/broadcast-and-unknown server
! and the client on atm subinterface 1/0.2. The configuration server assigns the
! client to the engineering emulated lan.
interface atm 1/0.2
 ip address 172.16.1.1 255.255.255.0
 lane server-bus tokenring eng
 lane client tokenring eng
```

Router 2

Router 2 is configured for a client of the manufacturing emulated LAN and a client of the engineering emulated LAN. Because the default emulated LAN name is *man*, the first client is linked to that emulated LAN name by default. Router 2 is configured as shown in this example:

```
interface atm 1/0
 atm pvc 1 0 5 qsaal
 atm pvc 2 0 16 ilmi
! This client is not in the configuration server's database, so it will be
! linked to the "man" ELAN by default.
interface atm 1/0.1
 ip address 172.16.0.2 255.255.255.0
 lane client tokenring
! A client for the following interface is entered in the configuration
! server's database as linked to the "eng" ELAN.
interface atm 1/0.2
 ip address 172.16.1.2 255.255.255.0
 lane client tokenring eng
```

Router 3

Router 3 is configured for a client of the manufacturing emulated LAN and a client of the marketing emulated LAN. Because the default emulated LAN name is *man*, the first client is linked to that emulated LAN name by default. The second client is listed in the database as linked to the *mkt* emulated LAN. Router 3 is configured as shown in this example:

```
interface atm 2/0
 atm pvc 1 0 5 qsaal
 atm pvc 2 0 16 ilmi
! The first client is not entered in the database, so it is linked to the
! "man" ELAN by default.
interface atm 2/0.1
 ip address 172.16.0.3 255.255.255.0
 lane client tokenring man
! The second client is explicitly entered in the configuration server's
! database as linked to the "mkt" ELAN.
interface atm 2/0.2
 ip address 172.16.2.3 255.255.255.0
 lane client tokenring mkt
```

Router 4

Router 4 has the server and BUS for the marketing emulated LAN, a client for marketing, and a client for manufacturing. The first client is listed in the database as linked to the *mkt* emulated LANs. The second client is not listed in the database, but is linked to the *man* emulated LAN name by default. Router 4 is configured as shown here:

```
interface atm 3/0
 atm pvc 1 0 5 qsaal
 atm pvc 2 0 16 ilmi
```

```
! The first client is explicitly entered in the configuration server's
! database as linked to the "mkt" ELAN.
interface atm 3/0.1
 ip address 172.16.2.4 255.255.255.0
 lane server-bus tokenring mkt
 lane client tokenring mkt
! The following client is not entered in the database, so it is linked to the
! "man" ELAN by default.
interface atm 3/0.2
 ip address 172.16.0.4 255.255.255.0
 lane client tokenring
```

TR-LANE with Two-Port Source-Route Bridging Example

The following example configures two Cisco 7500 series routers for one emulated Token Ring LAN using source-route bridging, as illustrated in Figure 21-3. This example does not restrict membership in the emulated LANs.

Figure 21-3 *Two-Port Source-Route Bridging TR-LANE*

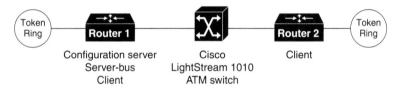

Router 1

Router 1 contains the configuration server, the server and BUS, and a client. Router 1 is configured as shown in this example:

```
hostname Router1
!
! The following lines configure the database cisco_eng.
lane database cisco_eng
 name elan1 server-atm-address 39.020304050607080910111213.00000CA05B41.01
 name elan1 local-seg-id 2048
 default-name elan1
!
interface Ethernet0/0
 ip address 10.6.10.4 255.255.255.0
!
! The following lines configure a configuration server using the cisco_eng database on
! the interface. No IP address is needed since we are using source-route bridging.
interface ATM2/0
 no ip address
 atm pvc 1 0 5 qsaal
 atm pvc 2 0 16 ilmi
 lane config auto-config-atm-address
 lane config database cisco_eng
!
```

```
! The following lines configure the server-bus and the client on the subinterface and
! specify source-route bridging information.
interface ATM2/0.1 multipoint
 lane server-bus tokenring elan1
 lane client tokenring elan1
 source-bridge 2048 1 1
 source-bridge spanning
!
! The following lines configure source-route bridging on the Token Ring interface.
interface TokenRing3/0/0
 no ip address
 ring-speed 16
 source-bridge 1 1 2048
 source-bridge spanning
!
router igrp 65529
 network 10.0.0.0
```

Router 2

Router 2 contains only a client for the emulated LAN. Router 2 is configured as shown here:

```
hostname Router2
!
interface Ethernet0/0
 ip address 10.6.10.5 255.255.255.0
!
! The following lines configure source-route bridging on the Token Ring interface.
interface TokenRing1/0
 no ip address
 ring-speed 16
 source-bridge 2 2 2048
 source-bridge spanning
!
! The following lines set up the signaling and ILMI PVCs.
interface ATM2/0
 no ip address
 atm pvc 1 0 5 qsaal
 atm pvc 2 0 16 ilmi
!
! The following lines set up a client on the subinterface and configure
! source-route bridging.
interface ATM2/0.1 multipoint
 ip address 1.1.1.2 255.0.0.0
 lane client tokenring elan1
 source-bridge 2048 2 2
 source-bridge spanning
!
router igrp 65529
 network 10.0.0.0
```

TR-LANE with Multiport Source-Route Bridging Example

The following example configures two Cisco 7500 series routers for one emulated Token-Ring LAN using source-route bridging, as illustrated in Figure 21-4. Since each router connects to three rings (the two Token Rings and the emulated LAN "ring"), a virtual ring must be configured on the router. This example does not restrict membership in the emulated LANs.

Figure 21-4 *Multiport Source-Route Bridged Token Ring Emulated LAN*

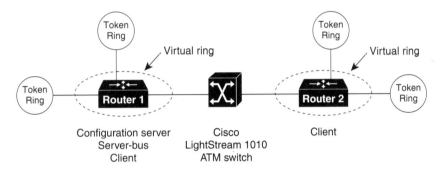

Router 1

Router 1 contains the configuration server, the server and BUS, and a client. Router 1 is configured as shown in this example:

```
hostname Router1
!
! The following lines configure the database with the information about the
! elan1 emulated Token Ring LAN.
lane database cisco_eng
 name elan1 server-atm-address 39.020304050607080910111213.00000CA05B41.01
 name elan1 local-seg-id 2048
 default-name elan1
!
! The following line configures virtual ring 256 on the router.
source-bridge ring-group 256
!
interface Ethernet0/0
 ip address 10.6.10.4 255.255.255.0
!
! The following lines configure the configuration server to use the cisco_eng database.
! The Signalling and ILMI PVCs are also configured.
interface ATM2/0
 no ip address
 atm pvc 1 0 5 qsaal
 atm pvc 2 0 16 ilmi
 lane config auto-config-atm-address
 lane config database cisco_eng
!
```

```
! The following lines configure the server and broadcast-and-unknown server and a client
! on the interface. The lines also specify source-route bridging information.
interface ATM2/0.1 multipoint
 lane server-bus tokenring elan1
 lane client tokenring elan1
 source-bridge 2048 5 256
 source-bridge spanning
!
! The following lines configure the Token Ring interfaces.
interface TokenRing3/0
 no ip address
 ring-speed 16
 source-bridge 1 1 256
 source-bridge spanning
interface TokenRing3/1
 no ip address
 ring-speed 16
 source-bridge 2 2 256
 source-bridge spanning
!
router igrp 65529
 network 10.0.0.0
```

Router 2

Router 2 contains only a client for the emulated LAN. Router 2 is configured as shown here:

```
hostname Router2
!
! The following line configures virtual ring 512 on the router.
source-bridge ring-group 512
!
interface Ethernet0/0
 ip address 10.6.10.5 255.255.255.0
!
! The following lines configure the Token Ring interfaces.
interface TokenRing1/0
 no ip address
 ring-speed 16
 source-bridge 3 3 512
 source-bridge spanning
interface TokenRing1/1
 no ip address
 ring-speed 16
 source-bridge 4 4 512
 source-bridge spanning
!
! The following lines configure the signaling and ILMI PVCs.
interface ATM2/0
 no ip address
 atm pvc 1 0 5 qsaal
 atm pvc 2 0 16 ilmi
!
! The following lines configure the client. Source-route bridging is also configured.
```

```
interface ATM2/0.1 multipoint
 ip address 1.1.1.2 255.0.0.0
 lane client tokenring elan1
 source-bridge 2048 6 512
 source-bridge spanning
!
router igrp 65529
 network 10.0.0.0
```

Routing Between Token Ring and Ethernet Emulated LANs Example

This example, shown in Figure 21-5, configures routing between a Token Ring emulated LAN (*trelan*) and an Ethernet emulated LAN (*ethelan*) on the same ATM interface. Router 1 contains the LANE configuration server, a LANE server and BUS for each emulated LAN, and a client for each emulated LAN. Router 2 contains a client for *trelan* (Token Ring); Router 3 contains a client for *ethelan* (Ethernet).

Figure 21-5 *Routing Between Token Ring and Ethernet Emulated LANs*

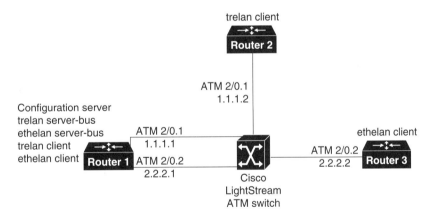

Router 1

Router 1 contains the LANE configuration server, a LANE server and BUS for each emulated LAN, and a client for each emulated LAN. Router 1 is configured as shown in this example:

```
hostname router1
!
! The following lines name and configure the configuration server's database.
! The server addresses for trelan and ethelan and the ELAN ring number for
! trelan are entered into the database. The default ELAN is trelan.
lane database cisco_eng
 name trelan server-atm-address 39.020304050607080910111213.00000CA05B41.01
 name trelan local-seg-id 2048
```

```
 name ethelan server-atm-address 39.020304050607080910111213.00000CA05B41.02
 default-name trelan
!
! The following lines enable the configuration server and associate it
! with the cisco_eng database.
interface ATM2/0
 no ip address
 atm pvc 1 0 5 qsaal
 atm pvc 2 0 16 ilmi
 lane config auto-config-atm-address
 lane config database cisco_eng
!
! The following lines configure the tokenring LES/BUS and LEC for trelan
! on subinterface atm2/0.1 and assign an IP address to the subinterface.
interface ATM2/0.1 multipoint
 ip address 10.1.1.1 255.255.255.0
 lane server-bus tokenring trelan
 lane client tokenring trelan
!
! The following lines configure the Ethernet LES/BUS and LEC for ethelan
! on subinterface atm2/0.2 and assign an IP address to the subinterface.
interface ATM2/0.2 multipoint
 ip address 20.2.2.1 255.255.255.0
 lane server-bus ethernet ethelan
 lane client ethernet ethelan
!
! The following lines configure the IGRP routing protocol to enable routing
! between ELANS.
router igrp 1
 network 10.0.0.0
 network 20.0.0.0
```

Router 2

Router 2 contains a client for *trelan* (Token Ring). Router 2 is configured as follows:

```
hostname router2
!
! The following lines set up the signaling and ILMI PVCs for the interface.
interface ATM2/0
 no ip address
 no keepalive
 atm pvc 1 0 5 qsaal
 atm pvc 2 0 16 ilmi
!
! The following lines configure a Token Ring LEC on atm2/0.1 and assign
! an IP address to the subinterface.
interface ATM2/0.1 multipoint
 ip address 10.1.1.2 255.255.255.0
 lane client tokenring trelan
!
! The following lines configure the IGRP routing protocol to enable routing
! between ELANS.
router igrp 1
```

```
network 10.0.0.0
network 20.0.0.0
```

Router 3

Router 3 contains a client for *ethelan* (Ethernet). Router 3 is configured as follows:

```
hostname router3
!
! The following lines set up the signaling and ILMI PVCs for the interface.
interface ATM2/0
 no ip address
 no ip mroute-cache
 atm pvc 1 0 5 qsaal
 atm pvc 2 0 16 ilmi
!
! The following lines configure an Ethernet LEC on atm2/0.1 and assign
! an IP address to the subinterface.
interface ATM2/0.1 multipoint
 ip address 20.2.2.2 255.255.255.0
 lane client ethernet ethelan
!
! The following lines configure the IGRP routing protocol to enable routing
! between ELANS.
router igrp 1
 network 10.0.0.0
 network 20.0.0.0
```

CHAPTER 22

LAN Emulation Commands

This chapter describes the commands available to configure LANE in Cisco 4500 series routers, Cisco 4700 routers, and Cisco 7500 routers that contain an AIP or an NP-1A ATM NPM and are connected to a Cisco ATM switch.

For LANE configuration information and examples, refer to Chapter 21, "Configuring LAN Emulation."

NOTE   Because some LANE commands are used often and others are used very rarely, the command descriptions identify the commands you are most likely to use. See the "Usage Guidelines" section for the following indicator: "This command is ordinarily used."

---

clear atm vc

To release a specified SVC, use the **clear atm vc** EXEC command.

> **clear atm vc** *vcd*

Syntax	Description
vcd	Virtual channel descriptor of the channel to be released.

Command Mode
EXEC

Usage Guidelines
This command first appeared in Cisco IOS Release 11.0.

For multicast or control VCCs, this command causes the LANE client to exit and rejoin an emulated LAN.

For data VCCs, this command also removes the associated LE ARP table entries.

Example
The following example releases SVC 1024:

```
clear atm vc 1024
```

clear lane le-arp

To clear the dynamic LE ARP table or a single LE ARP entry of the LANE client configured on the specified subinterface or emulated LAN, use the **clear lane le-arp** EXEC command.

> **clear lane le-arp** [**interface** *slot/port*[*.subinterface-number*] | **name** *elan-name*] [**mac-address** *mac-address* | **route-desc segment** *segment-number* **bridge** *bridge-number*] (for the Cisco 7500 series)

> **clear lane le-arp** [**interface** *number*[*.subinterface-number*] | **name** *elan-name*] [**mac-address** *mac-address* | **route-desc segment** *segment-number* **bridge** *bridge-number*]] (for the Cisco 4500 and 4700 routers)

Syntax / Description

Syntax	Description
interface *slot/port*[*.subinterface-number*]	(Optional) Interface or subinterface for the LANE client whose LE ARP table or entry is to be cleared for the Cisco 7500 series routers. The space between the **interface** keyword and the *slot* argument is optional.
interface *number*[*.subinterface-number*]	(Optional) Interface or subinterface for the LANE client whose LE ARP table or entry is to be cleared for the Cisco 4500 or 4700 routers. The space between the **interface** keyword and the *number* argument is optional.
name *elan-name*	(Optional) Name of the emulated LAN for the LANE client whose LE ARP table or entry is to be cleared. Maximum length is 32 characters.
mac-address *mac-address*	(Optional) MAC address of the entry to be cleared from the LE ARP table.
route-desc segment *segment-number*	(Optional) LANE segment number. The segment number ranges from 1 to 4,095.
bridge *bridge-number*	(Optional) Bridge number that is contained in the route descriptor. Valid bridge numbers range from 1 to 15.

Command Mode

EXEC

Usage Guidelines

This command first appeared in Cisco IOS Release 11.0.

This command removes dynamic LE ARP table entries only. It does not remove static LE ARP table entries.

If you do not specify an interface or an emulated LAN, this command clears all the LE ARP tables of any LANE client in the router.

If you specify a major interface (not a subinterface), this command clears all the LE ARP tables of every LANE client on all the subinterfaces of that interface.

This command also removes the fast-cache entries built from the LE ARP entries.

Examples

The following example clears all the LE ARP tables for all clients on the router:

```
clear lane le-arp
```

The following example clears all the LE ARP tables for all LANE clients on all the subinterfaces of interface 1/0:

```
clear lane le-arp interface 1/0
```

The following example clears the entry corresponding to MAC address 0800.AA00.0101 from the LE ARP table for the LANE client on the emulated LAN *red*:

```
clear lane le-arp name red 0800.aa00.0101
```

The following example clears all dynamic entries from the LE ARP table for the LANE client on the emulated LAN *red*:

```
clear lane le-arp name red
```

The following example clears the dynamic entry from the LE ARP table for the LANE client on segment number 1, bridge number 1 in the emulated LAN *red*:

```
clear lane le-arp name red route-desc segment 1 bridge 1
```

NOTE   MAC addresses are written in the same dotted notation for the **clear lane le-arp** command as they are for the global IP **arp** command.

---

clear lane server

To force a LANE server to drop a client and allow the LANE configuration server to assign the client to another emulated LAN, use the **clear lane server** EXEC command.

clear lane server {**interface** *slot/port* [*.subinterface-number*] | **name** *elan-name*}[**mac-address** *mac-address* | **client-atm-address** *atm-address* | **lecid** *lane-client-id* | **route-desc segment** *segment-number* **bridge** *bridgenumber*] (for the Cisco 7500 series)

clear lane server {**interface** *number* [*.subinterface-number*] | **name** *elan-name*} [**mac-address** *mac-address* | **client-atm-address** *atm-address* | **lecid** *lecid* | **route-desc segment** *segment-number* **bridge** *bridge-number*](for the Cisco 4500 and 4700 routers)

Syntax

Syntax	Description
interface *slot/port*[*.subinterface-number*]	Interface or subinterface where the LANE server is configured for the Cisco 7500 series. The space between the **interface** keyword and the *slot* argument is optional.
interface *number*[*.subinterface-number*]	Interface or subinterface where the LANE server is configured for the Cisco 4500 or 4700 routers. The space between the **interface** keyword and the *number* argument is optional.
name *elan-name*	Name of the emulated LAN on which the LANE server is configured. Maximum length is 32 characters.
mac-address *mac-address*	(Optional) Keyword and LANE client's MAC address.
client-atm-address *atm-address*	(Optional) Keyword and LANE client's ATM address.
lecid *lane-client-id*	(Optional) Keyword and LANE client ID. The LANE client ID is a value between 1 and 4,096.
route-desc segment *segment-number*	(Optional) Keywords and LANE segment number. The segment number ranges from 1 to 4,095.
bridge *bridge-number*	(Optional) Keyword and bridge number that is contained in the route descriptor. The bridge number ranges from 1 to 15.

Command Mode
EXEC

Usage Guidelines

This command first appeared in Cisco IOS Release 11.0.

After changing the bindings on the configuration server, use this command on the LANE server to force the client to leave one emulated LAN. The LANE server will drop the Control Direct and Control Distribute VCCs to the LANE client. The client will then ask the LANE configuration server for the location of the LANE server of the emulated LAN it should join.

If no LANE client is specified, all LANE clients attached to the LANE server are dropped.

Example

The following example forces all the LANE clients on the emulated LAN *red* to be dropped; the next time they try to join, they will be forced to join a different emulated LAN:

```
clear lane server name red
```

Related Commands

You can search online at www.cisco.com to find documentation of related commands.

client-atm-address name
lane database
mac-address name
show lane server

client-atm-address name

To add a LANE client address entry to the configuration server's configuration database, use the **client-atm-address** database configuration command. To remove a client address entry from the table, use the **no** form of this command.

> **client-atm-address** *atm-address-template* **name** *elan-name*
>
> **no client-atm-address** *atm-address-template*

Syntax	Description
atm-address-template	Template that explicitly specifies an ATM address or a specific part of an ATM address and uses wildcard characters for other parts of the ATM address, making it easy and convenient to specify multiple addresses matching the explicitly specified part. Wildcard characters can replace any nibble or group of nibbles in the prefix, the ESI, or the selector fields of the ATM address.
elan-name	Name of the emulated LAN. Maximum length is 32 characters.

Default

No address and no emulated LAN name are provided.

Command Mode

Database configuration

Usage Guidelines

This command first appeared in Cisco IOS Release 11.0.

This command is ordinarily used.

The effect of this command is to bind any client whose address matches the specified template into the specified emulated LAN. When a client comes up, it consults the LANE configuration server, which responds with the ATM address of the LANE server for the emulated LAN. The client then initiates join procedures with the LANE server.

Before this command is used, the emulated LAN specified by the *elan-name* argument must have been created in the configuration server's database by use of the **name server-atm-address** command.

If an existing entry in the configuration server's database binds the LANE client ATM address to a different emulated LAN, the new command is rejected.

This command affects only the bindings in the named configuration server database. It has no effect on the LANE components themselves.

See the **lane database** command for information about creating the database, and the **name server-atm-address** command for information about binding the emulated LAN's name to the server's ATM address.

The **client-atm-address name** command is a subcommand of the global **lane database** command.

ATM Addresses. A LANE ATM address has the same syntax as an NSAP, but it is not a network level address. It consists of the following:

* A 13-byte prefix that includes the following fields defined by the ATM Forum:

  — AFI (Authority and Format Identifier) field (1 byte), DCC (Data Country Code) or ICD (International Code Designator) field (2 bytes), DFI field (Domain Specific Part Format Identifier) (1 byte), Administrative Authority field (3 bytes), Reserved field (2 bytes), Routing Domain field (2 bytes), and the Area field (2 bytes)

* A 6-byte ESI

* A 1-byte selector field

Address Templates. LANE ATM address templates can use two types of wildcard characters: an asterisk (*) to match any single character (nibble), and an ellipsis (...) to match any number of leading, middle, or trailing characters. The values of the characters replaced by wildcards come from the automatically assigned ATM address.

In LANE, a *prefix template* explicitly matches the prefix but uses wildcards for the ESI and selector fields. An *ESI template* explicitly matches the ESI field but uses wildcards for the prefix and selector.

In Cisco's implementation of LANE, the prefix corresponds to the switch, the ESI corresponds to the ATM interface, and the selector field corresponds to the specific subinterface of the interface.

Examples

The following example uses an ESI template to specify the part of the ATM address corresponding to the interface. This example allows any client on any subinterface of the interface that corresponds to the displayed ESI value, no matter which switch the router is connected to, to join the *engineering* emulated LAN:

```
client-atm-address ...0800.200C.1001.** name engineering
```

The following example uses a prefix template to specify the part of the ATM address corresponding to the switch. This example allows any client on a subinterface of any interface connected to the switch that corresponds to the displayed prefix to join the *marketing* emulated LAN:

```
client-atm-address 47.000014155551212f.00.00... name marketing
```

Related Commands

You can search online at www.cisco.com to find documentation of related commands.

default-name
lane database
mac-address name
name server-atm-address

default-name

To provide an emulated LAN name in the configuration server's database for those client MAC addresses and client ATM addresses that do not have explicit emulated LAN name bindings, use the **default-name** database configuration command. To remove the default name, use the **no** form of this command.

> **default-name** *elan-name*

> **no default-name**

Syntax

elan-name

Description

Default emulated LAN name for any LANE client MAC address or LANE client ATM address not explicitly bound to any emulated LAN name. Maximum length is 32 characters.

Default

No name is provided.

Command Mode

Database configuration

Usage Guidelines

This command first appeared in Cisco IOS Release 11.0.

This command affects only the bindings in the configuration server's database. It has no effect on the LANE components themselves.

The named emulated LAN must already exist in the configuration server's database before this command is used. If the default name-to-emulated LAN name binding already exists, the new binding replaces it.

The **default-name** command is a subcommand of the global **lane database** command.

Example

The following example specifies the emulated Token Ring LAN *man* as the default emulated LAN. Because none of the emulated LANs are restricted, clients are assigned to whichever emulated LAN they request. Clients that do not request a particular emulated LAN will be assigned to the *man* emulated LAN.

```
lane database example2
 name eng server-atm-address 39.000001415555121101020304.0800.200c.1001.02
 name eng local-seg-id 1000
 name man server-atm-address 39.000001415555121101020304.0800.200c.1001.01
 name man local-seg-id 2000
 name mkt server-atm-address 39.000001415555121101020304.0800.200c.4001.01
 name mkt local-seg-id 3000
 default-name man
```

Related Commands

You can search online at www.cisco.com to find documentation of related commands.

client-atm-address name
lane database
mac-address name
name server-atm-address

lane auto-config-atm-address

To specify that the configuration server ATM address is computed by Cisco's automatic method, use the **lane auto-config-atm-address** interface configuration command. To remove the previously assigned ATM address, use the **no** form of this command.

> **lane [config] auto-config-atm-address**

> **no lane [config] auto-config-atm-address**

Syntax

Syntax	Description
config	(Optional) When the **config** keyword is used, this command applies only to the LECS. This keyword indicates that the LECS should use the auto-computed LECS address.

Default

No specific ATM address is set.

Command Mode

Interface configuration

Usage Guidelines

This command first appeared in Cisco IOS Release 11.0.

When the **config** keyword is not present, this command causes the LANE server and LANE client on the subinterface to use the automatically assigned ATM address for the configuration server.

When the **config** keyword is present, this command assigns the automatically generated ATM address to the LECS configured on the interface. Multiple commands that assign ATM addresses to the LANE configuration server can be issued on the same interface to assign different ATM addresses to the configuration server. Commands that assign ATM addresses to the LANE configuration server include **lane auto-config-atm-address**, **lane config-atm-address**, and **lane fixed-config-atm-address**.

For a discussion of Cisco's method of automatically assigning ATM addresses, refer to Chapter 21, "Configuring LAN Emulation."

Examples

The following example associates the LANE configuration server with the database named *network1* and specifies that the configuration server's ATM address will be assigned by Cisco's automatic method:

```
lane database network1
 name eng server-atm-address 39.020304050607080910111213.0800.AA00.1001.02
 name mkt server-atm-address 39.020304050607080910111213.0800.AA00.4001.01
interface atm 1/0
 lane config database network1
 lane config auto-config-atm-address
```

The following example causes the LANE server and LANE client on the subinterface to use the automatically assigned ATM address to communicate with the configuration server:

```
interface atm 2/0.1
 ip address 172.16.0.4 255.255.255.0
 lane client ethernet
 lane server-bus ethernet eng
 lane auto-config-atm-address
```

Related Commands

You can search online at www.cisco.com to find documentation of related commands.

lane config-atm-address
lane database
lane fixed-config-atm-address

lane bus-atm-address

To specify an ATM address—and thus override the automatic ATM address assignment—for the BUS on the specified subinterface, use the **lane bus-atm-address** interface configuration command. To remove the ATM address previously specified for the BUS on the specified subinterface, and thus, revert to the automatic address assignment, use the **no** form of this command.

> **lane bus-atm-address** *atm-address-template*
>
> **no lane bus-atm-address** [*atm-address-template*]

Syntax	Description
atm-address-template	ATM address or a template in which wildcard characters are replaced by any nibble or group of nibbles of the prefix bytes, the ESI bytes, or the selector byte of the automatically assigned ATM address.

Default

For the BUS, the default is automatic ATM address assignment.

Command Mode

Interface configuration

Usage Guidelines

This command first appeared in Cisco IOS Release 11.0.

When applied to a BUS, this command overrides automatic ATM address assignment for the BUS. When applied to a LANE client, this command gives the client the ATM address of the BUS. The client will use this address rather than sending LE ARP requests for the broadcast address.

When applied to a selected interface, but with a different ATM address than was used previously, this command replaces the BUS's ATM address.

ATM Addresses. A LANE ATM address has the same syntax as an NSAP (but it is not a network level address). It consists of the following:

- A 13-byte prefix that includes the following fields defined by the ATM Forum:
    - AFI field (1 byte)
    - DCC or ICD field (2 bytes)
    - DFI field (1 byte)
    - Administrative Authority field (3 bytes)
    - Reserved field (2 bytes)
    - Routing Domain field (2 bytes)
    - Area field (2 bytes)
- A 6-byte ESI
- A 1-byte selector field

Address Templates. LANE ATM address templates can use two types of wildcard characters: an asterisk (*) to match any single character (nibble), and an ellipsis (...) to match any number of leading, middle, or trailing characters. The values of the characters replaced by wildcards come from the automatically assigned ATM address.

The values of the digits that are replaced by wildcards come from the automatic ATM assignment method.

In LANE, a *prefix template* explicitly matches the prefix but uses wildcards for the ESI and selector fields. An *ESI template* explicitly matches the ESI field but uses wildcards for the prefix and selector.

In Cisco's implementation of LANE, the prefix corresponds to the switch, the ESI corresponds to the ATM interface, and the Selector field corresponds to the specific subinterface of the interface.

Examples

The following example uses an ESI template to specify the part of the ATM address corresponding to the interface; the remaining values in the ATM address come from automatic assignment:

```
lane bus-atm-address ...0800.200C.1001.**
```

The following example uses a prefix template to specify the part of the ATM address corresponding to the switch; the remaining values in the ATM address come from automatic assignment:

```
lane bus-atm-address 45.000014155551212f.00.00...
```

Related Commands

You can search online at www.cisco.com to find documentation of related commands.

lane server-bus

lane client

To activate a LANE client on the specified subinterface, use the **lane client** interface configuration command. To remove a previously activated LANE client on the subinterface, use the **no** form of this command.

> **lane client** {**ethernet** | **tokenring**} [*elan-name*]
>
> **no lane client** [{**ethernet** | **tokenring**} [*elan-name*]]

Syntax	Description
ethernet	Identifies the emulated LAN attached to this subinterface as an Ethernet emulated LAN.
tokenring	Identifies the emulated LAN attached to this subinterface as a Token Ring emulated LAN.
elan-name	(Optional) Name of the emulated LAN. This argument is optional because the client obtains its emulated LAN name from the configuration server. The maximum length of the name is 32 characters.

Default

No LANE clients are enabled on the interface.

Command Mode

Interface configuration

Usage Guidelines

This command first appeared in Cisco IOS Release 11.0.

If a **lane client** command has already been used on the subinterface for a different emulated LAN, then the client initiates termination procedures for that emulated LAN and joins the new emulated LAN.

If you do not provide an *elan-name* value, the client contacts the server to find which emulated LAN to join. If you do provide an emulated LAN name, the client consults the configuration server to ensure that no conflicting bindings exist.

Example

The following example enables a Token Ring LANE client on an interface:

```
lane client tokenring
```

Related Commands

You can search online at www.cisco.com to find documentation of related commands.

lane client-atm-address

lane client-atm-address

To specify an ATM address—and thus, override the automatic ATM address assignment—for the LANE client on the specified subinterface, use the **lane client-atm-address** interface configuration command. To remove the ATM address previously specified for the LANE client on the specified subinterface, and thus, revert to the automatic address assignment, use the **no** form of this command.

> **lane client-atm-address** *atm-address-template*
>
> **no client-atm-address** [*atm-address-template*]

Syntax	Description
atm-address-template	ATM address or a template in which wildcard characters are replaced by any nibble or group of nibbles of the prefix bytes, the ESI bytes, or the selector byte of the automatically assigned ATM address.

Default

Automatic ATM address assignment.

Command Mode

Interface configuration

Usage Guidelines

This command first appeared in Cisco IOS Release 11.0.

Use of this command on a selected subinterface, but with a different ATM address than was used previously, replaces the LANE client's ATM address.

ATM Addresses. A LANE ATM address has the same syntax as an NSAP (but it is not a network level address). It consists of the following:

- A 13-byte prefix that includes the following fields defined by the ATM Forum:
  - AFI field (1 byte)
  - DCC or ICD field (2 bytes)
  - DFI field (1 byte)
  - Administrative Authority field (3 bytes)
  - Reserved field (2 bytes)
  - Routing Domain field (2 bytes)
  - Area field (2 bytes)
- A 6-byte ESI
- A 1-byte selector field

Address Templates. LANE ATM address templates can use two types of wildcard characters: an asterisk (*) to match any single character (nibble), and an ellipsis (...) to match any number of leading, middle, or trailing characters. The values of the characters replaced by wildcards come from the automatically assigned ATM address.

In LANE, a *prefix template* explicitly matches the ATM address prefix but uses wildcards for the ESI and selector fields. An *ESI template* explicitly matches the ESI field but uses wildcards for the prefix and selector.

In Cisco's implementation of LANE, the prefix corresponds to the switch, the ESI corresponds to the ATM interface, and the selector field corresponds to the specific subinterface of the interface.

For a discussion of Cisco's method of automatically assigning ATM addresses, refer to Chapter 21, "Configuring LAN Emulation."

Examples

The following example uses an ESI template to specify the part of the ATM address corresponding to the interface; the remaining parts of the ATM address come from automatic assignment:

```
lane client-atm-address...0800.200C.1001.**
```

The following example uses a prefix template to specify the part of the ATM address corresponding to the switch; the remaining parts of the ATM address come from automatic assignment:

```
lane client-atm-address 47.000014155551212f.00.00...
```

Related Commands

You can search online at www.cisco.com to find documentation of related commands.

lane client

lane config-atm-address

To specify a configuration server's ATM address explicitly, use the **lane config-atm-address** interface configuration command. To remove an assigned ATM address, use the **no** form of this command.

> **lane [config] config-atm-address** *atm-address-template*

> **no lane [config] config-atm-address** *atm-address-template*

Syntax	Description
config	(Optional) When the **config** keyword is used, this command applies only to the LECS. This keyword indicates that the LECS should use the 20-byte address that you explicitly entered.
atm-address-template	ATM address or a template in which wildcard characters are replaced by any nibble or group of nibbles of the prefix bytes, the ESI bytes, or the selector byte of the automatically assigned ATM address.

Default

No specific ATM address or method is set.

Command Mode

Interface configuration

Usage Guidelines

This command first appeared in Cisco IOS Release 11.0.

If the **config** keyword is not present, this command causes the LANE server and LANE client on the subinterface to use the specified ATM address for the configuration server.

When the **config** keyword is present, this command adds an ATM address to the configuration server configured on the interface. A LANE configuration server can listen on multiple ATM addresses. Multiple commands that assign ATM addresses to the LANE configuration server can be issued on the same interface to assign different ATM addresses to the LANE configuration server.

ATM Addresses. A LANE ATM address has the same syntax as an NSAP (but it is not a network level address). It consists of the following:

- A 13-byte prefix that includes the following fields defined by the ATM Forum:

    — AFI field (1 byte)

    — DCC or ICD field (2 bytes)

    — DFI field (1 byte)

    — Administrative Authority field (3 bytes)

    — Reserved field (2 bytes)

    — Routing Domain field (2 bytes)

    — Area field (2 bytes)

- A 6-byte ESI

- A 1-byte selector field

Address Templates. LANE ATM address templates can use two types of wildcard characters: an asterisk (*) to match any single character (nibble), and an ellipsis (...) to match any number of leading, middle, or trailing characters. The values of the characters replaced by wildcards come from the automatically assigned ATM address.

In LANE, a *prefix template* explicitly matches the ATM address prefix but uses wildcards for the ESI and selector fields. An *ESI template* explicitly matches the ESI field but uses wildcards for the prefix and selector.

In Cisco's implementation of LANE, the prefix corresponds to the switch prefix, the ESI corresponds to a function of ATM interface's MAC address, and the Selector field corresponds to the specific subinterface of the interface.

For a discussion of Cisco's method of automatically assigning ATM addresses, refer to Chapter 21, "Configuring LAN Emulation."

Examples

The following example associates the LANE configuration server with the database named *network1* and explicitly specifies the configuration server's ATM address:

```
lane database network1
 name eng server-atm-address 39.020304050607080910111213.0800.AA00.1001.02
 name mkt server-atm-address 39.020304050607080910111213.0800.AA00.4001.01
interface atm 1/0
 lane config database network1
 lane config config-atm-address 39.020304050607080910111213.0800.AA00.3000.00
```

The following example causes the LANE server and LANE client on the subinterface to use the explicitly specified ATM address to communicate with the configuration server:

```
interface atm 2/0.1
 ip address 172.16.0.4 255.255.255.0
 lane client ethernet
 lane server-bus ethernet eng
 lane config-atm-address 39.020304050607080910111213.0800.AA00.3000.00
```

Related Commands

You can search online at www.cisco.com to find documentation of related commands.

lane auto-config-atm-address
lane config database
lane database
lane fixed-config-atm-address

lane config database

To associate a named configuration table (database) with the configuration server on the selected ATM interface, use the **lane config database** interface configuration command. To remove the association between a named database and the configuration server on the specified interface, use the **no** form of this command.

> **lane config database** *database-name*
>
> **no lane config**

Syntax	Description
database-name	Name of the LANE database.

Part
VII

Command Reference

Default

No configuration server is defined, and no database name is provided.

Command Mode

Interface configuration

Usage Guidelines

This command first appeared in Cisco IOS Release 11.0.

This command is valid only on a major interface, not a subinterface, because only one LANE configuration server can exist for per interface.

The named database must exist before the **lane config database** command is used. Refer to the **lane database** command for more information.

Multiple **lane config database** commands cannot be used multiple times on the same interface. You must delete an existing association by using the **no** form of this command before you can create a new association on the specified interface.

Activating a LANE configuration server requires the **lane config database** command and one of the following commands: **lane config fixed-config-atm-address**, **lane config auto-config-atm-address**, or **lane config config-atm-address**.

Example

The following example associates the LANE configuration server with the database named *network1* and specifies that the configuration server's ATM address will be assigned by Cisco's automatic method:

```
lane database network1
 name eng server-atm-address 39.020304050607080910111213.0800.AA00.1001.02
 name mkt server-atm-address 39.020304050607080910111213.0800.AA00.4001.01
interface atm 1/0
 lane config database network1
 lane config auto-config-atm-address
```

Related Commands

You can search online at www.cisco.com to find documentation of related commands.

lane auto-config-atm-address
lane config-atm-address
lane database
lane fixed-config-atm-address

lane database

To create a named configuration database that can be associated with a configuration server, use the **lane database** global configuration command. To delete the database, use the **no** form of this command.

> **lane database** *database-name*
>
> **no lane database** *database-name*

Syntax	Description
database-name	Database name (32 characters maximum).

Default

No name is provided.

Command Mode

Global configuration

Usage Guidelines

This command first appeared in Cisco IOS Release 11.0.

Use of the **lane database** command places you in database configuration mode, in which you can use the **client-atm-address name**, **default name**, **mac-address name**, **name restricted, name unrestricted, name new-name,** and **name server-atm-address** commands to create entries in the specified database. When you are finished creating entries, type **^Z** or **exit** to return to global configuration mode.

Example

The following example creates the database named *network1* and associates it with the configuration server on interface ATM 1/0:

```
lane database network1
 name eng server-atm-address 39.020304050607080910111213.0800.AA00.1001.02
 name mkt server-atm-address 39.020304050607080910111213.0800.AA00.4001.01
 default-name eng
interface atm 1/0
 lane config database network1
 lane config auto-config-atm-address
```

Related Commands

You can search online at www.cisco.com to find documentation of related commands.

client-atm-address name
default-name
lane config database
mac-address name
name new-name
name server-atm-address

lane fixed-config-atm-address

To specify that the fixed configuration server ATM address assigned by the ATM Forum will be used, use the **lane fixed-config-atm-address** interface configuration command. To specify that the fixed ATM address is not used, use the **no** form of this command.

> **lane [config] fixed-config-atm-address**

> **no lane [config] fixed-config-atm-address**

Syntax   Description

config   (Optional) When the **config** keyword is used, this command applies only
                  to the LECS. This keyword indicates that LECS should use the well-
                  known ATM Forum LEC address.

Default

No specific ATM address or method is set.

Command Mode

Interface configuration

Usage Guidelines

This command first appeared in Cisco IOS Release 11.0.

When the **config** keyword is not present, this command causes the LANE server and LANE client on the subinterface to use that ATM address, rather than the ATM address provided by the ILMI, to locate the configuration server.

When the **config** keyword is present, and the LECS is already up and running, be aware of the following scenarios:

- If you configure the LECS with only the well-known address, the LECS will not participate in the SSRP, act as a "standalone" master, and only listen on the well-known LECS address. This scenario is ideal if you want a standalone LECS that does not participate in SSRP, and you would like to listen to only the well-known address.

- If only the well-known address is already assigned, and you assign at least one other address to the LECS, (additional addresses are assigned using the **lane config auto-config-atm-address** command and/or the **lane config config-atm-address command**) the LECS will participate in the SSRP and act as the master or slave based on the normal SSRP rules. This scenario is ideal if you would like the LECS to participate in SSRP, and you would like to make the master LECS listen on the well-known address.

- If the LECS is participating in SSRP, has more than one address (one of which is the well-known address), and all the addresses but the well-known address is removed, the LECS will declare itself the master and stop participating in SSRP completely.

- If the LECS is operating as an SSRP slave, and it has the well-known address configured, it will not listen on the well-known address unless it becomes the master.

- If you want the LECS to assume the well-known address only when it becomes the master, configure the LECS with the well-known address and at least one other address.

When you use this command with the **config** keyword, and the LECS is a master, the master will listen on the fixed address. If you use this command when an LECS is not a master, the LECS will listen on this address when it becomes a master. If you do not use this command, the LECS will not listen on the fixed address.

Multiple commands that assign ATM addresses to the LECS can be issued on the same interface in order to assign different ATM addresses to the LECS. Commands that assign ATM addresses to the LECS include **lane auto-config-atm-address**, **lane config-atm-address**, and **lane fixed-config-atm-address**. The **lane config database** command and at least one command that assigns an ATM address to the LECS are required to activate a LECS.

Examples

The following example associates the LANE configuration server with the database named *network1* and specifies that the configuration server's ATM address is the fixed address:

```
lane database network1
 name eng server-atm-address 39.020304050607080910111213.0800.AA00.1001.02
 name mkt server-atm-address 39.020304050607080910111213.0800.AA00.4001.01
interface atm 1/0
 lane config database network1
 lane config fixed-config-atm-address
```

The following example causes the LANE server and LANE client on the subinterface to use the fixed ATM address to communicate with the configuration server:

```
interface atm 2/0.1
```

```
ip address 172.16.0.4 255.255.255.0
lane client ethernet
lane server-bus ethernet eng
lane fixed-config-atm-address
```

Related Commands

You can search online at www.cisco.com to find documentation of related commands.

Table lane auto-config-atm-address
lane config-atm-address
lane config database

lane global-lecs-address

To specify a list of LECS addresses to use when the addresses cannot be obtained from the ILMI, use the **lane global-lecs-address** interface configuration command. The **no** form of this command removes an LECS address from the list.

> **lane global-lecs-address** *address*
>
> **no lane global-lecs-address** *address*

Syntax	Description
address	Address of the LECS. You cannot use the well-known LECS address.

Default

No addresses are configured. The router obtains LECS addresses from the ILMI.

Command Mode

Interface configuration

Usage Guidelines

This command first appeared in Cisco IOS Release 11.2.

Use this command when your ATM switches do not support the ILMI list of LECS addresses and you want to configure Simple Server Redundancy. This command will simulate the list of LECS addresses, as if they were obtained from the ILMI. Use this command with a different address for each LECS. The order they are used determines their priority. You should enter the addresses in the same order as you would on the ATM switch.

NOTE	You must configure the same list of addresses on each interface that contains a LANE entity.

If your switches do support ILMI, this command force the router to use the addresses specified and will not use the ILMI to obtain the LECS addresses.

Since the well-known LECS address is always used as a last-resort LECS address, you cannot use the address in this command.

lane le-arp

To add a static entry to the LE ARP table of the LANE client configured on the specified subinterface, use the **lane le-arp** interface configuration command. To remove a static entry from the LE ARP table of the LANE client on the specified subinterface, use the **no** form of this command.

> **lane le-arp** {*mac-address* | **route-desc segment** *segment-number* **bridge** *bridge-number*}*atm-address*

> **no lane le-arp** {*mac-address* | **route-desc segment** *segment-number* **bridge** *bridge-number*}*atm-address*

Syntax	Description
mac-address	MAC address to bind to the specified ATM address.
segment-number	LANE segment number. The segment number ranges from 1 to 4,095.
bridge-number	Bridge number that is contained in the route descriptor. The bridge number ranges from 1 to 15.
atm-address	ATM address.

Default
No static address bindings are provided.

Command Mode
Interface configuration

Usage Guidelines

This command first appeared in Cisco IOS Release 11.0.

This command adds or removes a static entry binding a MAC address or segment number and bridge number to an ATM address. It does not add or remove dynamic entries. Removing the static entry for a specified ATM address from an LE ARP table does not release Data Direct VCCs established to that ATM address. However, clearing a static entry clears any fast-cache entries that were created from the MAC address-to-ATM address binding.

Static LE ARP entries are neither aged nor removed automatically.

To remove dynamic entries from the LE ARP table of the LANE client on the specified subinterface, use the **clear lane le-arp** command.

Examples

The following example adds a static entry to the LE ARP table:

```
lane le-arp 0800.aa00.0101 47.000014155551212f.00.00.0800.200C.1001.01
```

The following example adds a static entry to the LE ARP table binding segment number 1, bridge number 1 to the ATM address:

```
lane le-arp route-desc segment 1 bridge 1 39.020304050607080910111213.00000CA05B41.01
```

Related Commands

You can search online at www.cisco.com to find documentation of related commands.

clear lane le-arp

lane server-atm-address

To specify an ATM address—and thus override the automatic ATM address assignment—for the LANE server on the specified subinterface, use the **lane server-atm-address** interface configuration command. To remove the ATM address previously specified for the LANE server on the specified subinterface and thus revert to the automatic address assignment, use the **no** form of this command.

> **lane server-atm-address** *atm-address-template*
>
> **no server-atm-address** [*atm-address-template*]

Syntax	Description
atm-address-template	ATM address or a template in which wildcard characters are replaced by any nibble or group of nibbles of the prefix bytes, the ESI bytes, or the selector byte of the automatically assigned ATM address.

Default

For the LANE server, the default is automatic address assignment; the LANE client finds the LANE server by consulting the configuration server.

Command Mode

Interface configuration

Usage Guidelines

This command first appeared in Cisco IOS Release 11.0.

This command also instructs the LANE client configured on this subinterface to reach the LANE server by using the specified ATM address instead of the ATM address provided by the configuration server.

When used on a selected subinterface, but with a different ATM address than was used previously, this command replaces the LANE server's ATM address.

ATM Addresses. A LANE ATM address has the same syntax as an NSAP (but it is not a network level address). It consists of the following:

- A 13-byte prefix that includes the following fields defined by the ATM Forum:
    - AFI field (1 byte)
    - DCC or ICD field (2 bytes)
    - DFI field (1 byte)
    - Administrative Authority field (3 bytes)
    - Reserved field (2 bytes)
    - Routing Domain field (2 bytes)
    - Area field (2 bytes)
- A 6-byte ESI
- A 1-byte selector field

Address Templates. LANE ATM address templates can use two types of wildcard characters: an asterisk (*) to match any single character (nibble), and an ellipsis (...) to match any number of leading, middle, or trailing characters. The values of the characters replaced by wildcards come from the automatically assigned ATM address.

In LANE, a *prefix template* explicitly matches the prefix, but uses wildcards for the ESI and selector fields. An *ESI template* explicitly matches the ESI field, but uses wildcards for the prefix and selector.

In Cisco's implementation of LANE, the prefix corresponds to the switch, the ESI corresponds to the ATM interface, and the Selector field corresponds to the specific subinterface of the interface.

For a discussion of Cisco's method of automatically assigning ATM addresses, refer to Chapter 21, "Configuring LAN Emulation."

Examples

The following example uses an ESI template to specify the part of the ATM address corresponding to the interface; the remaining parts of the ATM address come from automatic assignment:

```
lane server-atm-address ...0800.200C.1001.**
```

The following example uses a prefix template to specify the part of the ATM address corresponding to the switch; the remaining part of the ATM address come from automatic assignment:

```
lane server-atm-address 45.000014155551212f.00.00...
```

Related Commands

You can search online at www.cisco.com to find documentation of related commands.

lane server-bus

lane server-bus

To enable a LANE server and a BUS on the specified subinterface with the ELAN ID, use the **lane server-bus** interface configuration command. To disable a LANE server and BUS on the specified subinterface, use the **no** form of this command.

lane server-bus {**ethernet** I **tokenring**} *elan-name* [**elan-id** *id*]

no lane server-bus {**ethernet** I **tokenring**} *elan-name* [**elan-id** *id*]

Syntax	Description
ethernet	Identifies the emulated LAN attached to this subinterface as an Ethernet emulated LAN.
tokenring	Identifies the emulated LAN attached to this subinterface as a Token Ring emulated LAN.
elan-name	Name of the emulated LAN. The maximum length of the name is 32 characters.
elan-id	Identifies the emulated LAN.
id	Specifies the ELAN ID of the LEC.

Default

No LAN type or emulated LAN name is provided.

Command Mode

Interface configuration

Usage Guidelines

This command first appeared in Cisco IOS Release 11.0.

The **elan-id** keyword first appeared in Cisco IOS Release 12.0.

The LANE server and the BUS are located on the same router.

If a **lane server-bus** command has already been used on the subinterface for a different emulated LAN, the server initiates termination procedures with all clients and comes up as the server for the new emulated LAN.

To participate in MPOA, a LEC must have an ELAN ID. This command enables the LEC to get the ELAN ID from the LES when the LEC bypasses the LECS phase.

CAUTION If an ELAN ID is supplied, make sure that it corresponds to the same ELAN ID value specified in the LECS for the same emulated LAN.

The LEC can also obtain the ELAN ID from the LECS by using the **name elan-id** command.

Example

The following example enables a LANE server and BUS for a Token Ring emulated LAN named *MYELAN*:

```
lane server-bus tokenring MYELAN
```

Related Commands

You can search online at www.cisco.com to find documentation of related commands.

lane server-atm-address
name elan-id

Part VII

Command Reference

name elan-id

To configure the ELAN ID of an emulated LAN in the LECS database to participate in MPOA, use the **name elan-id** LANE database configuration command. To disable, use the **no** form of this command.

> **name** *name* **elan-id** *id*
>
> **no name** *name* **elan-id** *id*

Syntax	Description
name	Specifies the name of the emulated LAN.
id	Specifies the identification number of the emulated LAN.

Default

No ELAN ID is configured.

Command Mode

LANE database configuration

Usage Guidelines

This command first appeared in Cisco IOS Release 12.0.

To participate in MPOA, a LEC must have an ELAN ID. The LEC obtains the ELAN ID from the LECS. In case the LEC bypasses the LECS phase, the LEC can get the ELAN ID from the LES by using the **name elan-id** command.

Example

The following example sets the ELAN ID to 10 for emulated LAN named *MYELAN*:

```
name MYELAN elan-id 10
```

Related Commands

You can search online at www.cisco.com to find documentation of related commands.

lane server-bus

name preempt

To set the emulated LAN preempt, use the **name preempt** LANE database configuration command. To disable preempt, use the **no** form of this command.

> **name** *elan-name* **preempt**
>
> **no name** *elan-name* **preempt**

Syntax	Description
elan-name	Specifies the name of the emulated LAN.

Default

Preempt is off by default.

Command Mode

LANE database configuration

Usage Guidelines

This command first appeared in Cisco IOS Release 11.3.

Previously, when the primary LES failed, Cisco SSRP switched over to a secondary LES. But when a LES that is ranked higher in the list came back up, SSRP switched the active LES to the new LES, which had a higher priority. This forced the network to flap multiple times. This release prevents the network flapping by staying with the currently active master LES regardless of the priority. If a higher-priority LES comes back online, SSRP does not switch to that LES.

LES preemption is off by default. The first LES that comes on becomes the master. Users can revert to the old behavior (of switching to the higher priority LES all the time) by specifying the **name** *elan-name* **preempt** command in the LECS database.

Example

The following example sets the emulated LAN preempt for the emulated LAN named *MYELAN*:

```
name MYELAN preempt
```

name local-seg-id

To specify or replace the ring number of the emulated LAN in the configuration server's configuration database, use the **name local-seg-id** database configuration command. To remove the ring number from the database, use the **no** form of this command.

> **name** *elan-name* **local-seg-id** *segment-number*
>
> **no name** *elan-name* **local-seg-id** *segment-number*

Syntax	Description
elan-name	Name of the emulated LAN. The maximum length of the name is 32 characters.
segment-number	Segment number to be assigned to the emulated LAN. The number ranges from 1 to 4,095.

Default

No emulated LAN name or segment number is provided.

Command Mode

Database configuration

Usage Guidelines

This command first appeared in Cisco IOS Release 11.3.

This command is ordinarily used for Token Ring LANE.

The same LANE ring number cannot be assigned to more than one emulated LAN.

The **no** form of this command deletes the relationships.

Example

The following example specifies a ring number of 1024 for the emulated LAN *red*:

```
name red local-seg-id 1024
```

Related Commands

You can search online at www.cisco.com to find documentation of related commands.

default-name
lane database
mac-address name

name server-atm-address

To specify or replace the ATM address of the LANE server for the emulated LAN in the configuration server's configuration database, use the **name server-atm-address** database configuration command. To remove it from the database, use the **no** form of this command.

> **name** *elan-name* **server-atm-address** *atm-address* [**restricted** | **un-restricted**] [**index** *number*]

> **no name** *elan-name* **server-atm-address** *atm-address* [**restricted** | **un-restricted**] [**index** *number*]

Syntax	Description
elan-name	Name of the emulated LAN. Maximum length is 32 characters.
atm-address	LANE server's ATM address.
restricted | **un-restricted**	(Optional) Membership in the named emulated LAN is restricted to the LANE clients explicitly defined to the emulated LAN in the configuration server's database.
index *number*	(Optional) Priority number. When specifying multiple LANE servers for fault tolerance, you can specify a priority for each server. 0 is the highest priority.

Default

No emulated LAN name or server ATM address are provided.

Command Mode

Database configuration

Usage Guidelines

This command first appeared in Cisco IOS Release 11.0. The **restricted** command first appeared in Cisco IOS Release 11.1. The **un-restricted** and **index** keywords first appeared in Cisco IOS Release 11.2.

Emulated LAN names must be unique within one named LANE configuration database.

Specifying an existing emulated LAN name with a new LANE server ATM address adds the LANE server ATM address for that emulated LAN for redundant server operation or simple LANE service replication. This command can be used multiple times.

The **no** form of this command deletes the relationships.

Example

The following example configures the *example3* database with two restricted and one unrestricted emulated LANs. The clients that can be assigned to the *eng* and *mkt* emulated LANs are specified using the **client-atm-address** commands. All other clients are assigned to the *man* emulated LAN.

```
lane database example3
 name eng server-atm-address 39.000001415555121101020304.0800.200c.1001.02 restricted
 name man server-atm-address 39.000001415555121101020304.0800.200c.1001.01
 name mkt server-atm-address 39.000001415555121101020304.0800.200c.4001.01 restricted
 client-atm-address 39.000001415555121101020304.0800.200c.1000.02 name eng
 client-atm-address 39.000001415555121101020304.0800.200c.2000.02 name eng
 client-atm-address 39.000001415555121101020304.0800.200c.3000.02 name mkt
 client-atm-address 39.000001415555121101020304.0800.200c.4000.01 name mkt
 default-name man
```

Related Commands

You can search online at www.cisco.com to find documentation of related commands.

client-atm-address name
default-name
lane database
mac-address name

show lane

To display detailed information for all the LANE components configured on an interface or any of its subinterfaces, on a specified subinterface, or on an emulated LAN, use the **show lane** EXEC command.

> **show lane** [**interface atm** *slot/port*[*.subinterface-number*] | **name** *elan-name*] [**brief**]
> (for the AIP on the Cisco 7500 series routers; for the ATM port adapter on the Cisco 7200 series)

show lane [**interface atm** *slot/port-adapter/port*[*.subinterface-number*] | **name** *elan-name*][**brief**] (for the ATM port adapter on the Cisco 7500 series routers)

show lane [**interface atm** *number*[*.subinterface-number*] | **name** *elan-name*] [**brief**] (for the Cisco 4500 and 4700 routers)

Syntax	Description
interface atm *slot/port*	(Optional) ATM interface slot and port for the following:
	• AIP on the Cisco 7500 series routers.
	• ATM port adapter on the Cisco 7200 series routers.
interface atm *slot/port-adapter/port*	(Optional) ATM interface slot, port adapter, and port number for the ATM port adapter on the Cisco 7500 series routers.
interface atm *number*	(Optional) ATM interface number for the NPM on the Cisco 4500 or 4700 routers.
.subinterface-number	(Optional) Subinterface number.
name *elan-name*	(Optional) Name of emulated LAN. The maximum length of the name is 32 characters.
brief	(Optional) Keyword used to display the brief subset of available information.

Command Mode
EXEC

Usage Guidelines
This command first appeared in Cisco IOS Release 11.0.

Using the **show lane** command is equivalent to using the **show lane config**, **show lane server**, **show lane bus**, and **show lane client** commands. The **show lane** command shows all LANE-related information except the **show lane database** command information.

Sample Displays

The following is sample output from the **show lane** command for an Ethernet emulated LAN:

```
Router# show lane

LE Config Server ATM2/0 config table: cisco_eng
Admin: up  State: operational
LECS Mastership State: active master
list of global LECS addresses (30 seconds to update):
39.020304050607080910111213.00000CA05B43.00  <-------- me
ATM Address of this LECS: 39.020304050607080910111213.00000CA05B43.00 (auto)
 vcd  rxCnt txCnt  callingParty
 50     2     2    39.020304050607080910111213.00000CA05B41.02 LES elan2 0 active
cumulative total number of unrecognized packets received so far: 0
cumulative total number of config requests received so far: 30
cumulative total number of config failures so far: 12
    cause of last failure: no configuration
    culprit for the last failure: 39.020304050607080910111213.00602F557940.01


LE Server ATM2/0.2  ELAN name: elan2 Admin: up  State: operational
type: ethernet         Max Frame Size: 1516
ATM address: 39.020304050607080910111213.00000CA05B41.02
LECS used: 39.020304050607080910111213.00000CA05B43.00 connected, vcd 51
control distribute: vcd 57, 2 members, 2 packets

proxy/ (ST: Init, Conn, Waiting, Adding, Joined, Operational, Reject, Term)
lecid ST vcd    pkts Hardware Addr  ATM Address
  1  0  54       2 0000.0ca0.5b40 39.020304050607080910111213.00000CA05B40.02
  2  0  81       2 0060.2f55.7940 39.020304050607080910111213.00602F557940.02

LE BUS ATM2/0.2  ELAN name: elan2  Admin: up  State: operational
type: ethernet         Max Frame Size: 1516
ATM address: 39.020304050607080910111213.00000CA05B42.02
data forward: vcd 61, 2 members, 0 packets, 0 unicasts

lecid  vcd    pkts  ATM Address
  1    58       0 39.020304050607080910111213.00000CA05B40.02
  2    82       0 39.020304050607080910111213.00602F557940.02

LE Client ATM2/0.2  ELAN name: elan2  Admin: up  State: operational
Client ID: 1               LEC up for 11 minutes 49 seconds
Join Attempt: 1
HW Address: 0000.0ca0.5b40  Type: ethernet         Max Frame Size: 1516

ATM Address: 39.020304050607080910111213.00000CA05B40.02

 VCD  rxFrames  txFrames  Type      ATM Address
  0      0         0     configure  39.020304050607080910111213.00000CA05B43.00
  55     1         4     direct     39.020304050607080910111213.00000CA05B41.02
  56     6         0     distribute 39.020304050607080910111213.00000CA05B41.02
  59     0         1     send       39.020304050607080910111213.00000CA05B42.02
  60     3         0     forward    39.020304050607080910111213.00000CA05B42.02
  84     3         5     data       39.020304050607080910111213.00602F557940.02
```

The following is sample output from the **show lane** command for a Token Ring LANE network:

```
Router# show lane

LE Config Server ATM4/0 config table: eng
Admin: up  State: operational
LECS Mastership State: active master
list of global LECS addresses (35 seconds to update):
39.020304050607080910111213.006047704183.00  <-------- me
ATM Address of this LECS: 39.020304050607080910111213.006047704183.00 (auto)
 vcd  rxCnt  txCnt  callingParty
   7     1      1   39.020304050607080910111213.006047704181.01 LES elan1 0 active
cumulative total number of unrecognized packets received so far: 0
cumulative total number of config requests received so far: 2
cumulative total number of config failures so far: 0


LE Server ATM4/0.1  ELAN name: elan1  Admin: up  State: operational
type: token ring       Max Frame Size: 4544     Segment ID: 2048
ATM address: 39.020304050607080910111213.006047704181.01
LECS used: 39.020304050607080910111213.006047704183.00 connected, vcd 9
control distribute: vcd 12, 1 members, 2 packets

proxy/ (ST: Init, Conn, Waiting, Adding, Joined, Operational, Reject, Term)
lecid ST vcd    pkts Hardware Addr  ATM Address
   1  0   8      3 100.2          39.020304050607080910111213.006047704180.01
                   0060.4770.4180 39.020304050607080910111213.006047704180.01


LE BUS ATM4/0.1  ELAN name: elan1  Admin: up  State: operational
type: token ring       Max Frame Size: 4544     Segment ID: 2048
ATM address: 39.020304050607080910111213.006047704182.01
data forward: vcd 16, 1 members, 0 packets, 0 unicasts

lecid  vcd    pkts  ATM Address
   1   13       0 39.020304050607080910111213.006047704180.01


LE Client ATM4/0.1  ELAN name: elan1  Admin: up  State: operational
Client ID: 1               LEC up for 2 hours 25 minutes 39 seconds
Join Attempt: 3
HW Address: 0060.4770.4180   Type: token ring       Max Frame Size: 4544
Ring:100    Bridge:2        ELAN Segment ID: 2048
ATM Address: 39.020304050607080910111213.006047704180.01

 VCD  rxFrames  txFrames  Type      ATM Address
   0      0        0  configure  39.020304050607080910111213.006047704183.00
  10      1        3  direct     39.020304050607080910111213.006047704181.01
  11      2        0  distribute 39.020304050607080910111213.006047704181.01
  14      0        0  send       39.020304050607080910111213.006047704182.01
  15      0        0  forward    39.020304050607080910111213.006047704182.01
```

Table 22-1 describes significant fields in the sample displays of the **show lane** command.

Table 22-1   *show lane Field Descriptions*

Field	Description
LE Config Server	Identifies the following lines as applying to the LANE configuration server. These lines are also displayed in output from the **show lane config** command. See the **show lane config** command for explanations of the output.
LE Server	Identifies the following lines as applying to the LANE server. These lines are also displayed in output from the **show lane server** command. See the **show lane server** command for explanations of the output.
LE BUS	Identifies the following lines as applying to the LANE BUS. These lines are also displayed in output from the **show lane bus** command. See the **show lane bus** command for explanations of the output.
LE Client	Identifies the following lines as applying to a LANE client. These lines are also displayed in output from the **show lane client** command. See the **show lane bus** command for explanations of the output.

show lane bus

To display detailed LANE information for the BUS configured on an interface or any of its subinterfaces, on a specified subinterface, or on an emulated LAN, use the **show lane bus** EXEC command:

> **show lane bus** [**interface atm** *slot/port*[*.subinterface-number*] | **name** *elan-name*] [**brief**] (for the AIP on the Cisco 7500 series routers; for the ATM port adapter on the Cisco 7200 series)

> **show lane bus** [**interface atm** *slot/port-adapter/port*[*.subinterface-number*] | **name** *elan-name*] [**brief**] (for the ATM port adapter on the Cisco 7500 series routers)

> **show lane bus** [**interface atm** *number*[*.subinterface-number*] | **name** *elan-name*] [**brief**] (for the Cisco 4500 and 4700 routers)

Syntax	Description
interface atm *slot/port*	(Optional) ATM interface slot and port for the following: • AIP on the Cisco 7500 series routers. • ATM port adapter on the Cisco 7200 series routers.
interface atm *slot/port-adapter/port*	(Optional) ATM interface slot, port adapter, and port number for the ATM port adapter on the Cisco 7500 series routers.

interface atm *number*

(Optional) ATM interface number for the NPM on the Cisco 4500 or 4700 routers.

.subinterface-number

(Optional) Subinterface number.

name *elan-name*

(Optional) Name of emulated LAN. The maximum length of the name is 32 characters.

brief

(Optional) Keyword used to display the brief subset of available information.

Command Mode
EXEC

Usage Guidelines
This command first appeared in Cisco IOS Release 11.0.

Sample Displays
The following is sample output from the **show lane bus** command for an Ethernet emulated LAN:

```
Router# show lane bus

LE BUS ATM2/0.2  ELAN name: elan2  Admin: up  State: operational
type: ethernet        Max Frame Size: 1516
ATM address: 39.020304050607080910111213.00000CA05B42.02
data forward: vcd 61, 2 members, 0 packets, 0 unicasts

lecid  vcd    pkts   ATM Address
   1   58        0 39.020304050607080910111213.00000CA05B40.02
   2   82        0 39.020304050607080910111213.00602F557940.02
```

The following is sample output from the **show lane bus** command for a Token Ring LANE:

```
Router# show lane bus

LE BUS ATM3/0.1  ELAN name: anubis  Admin: up  State: operational
type: token ring       Max Frame Size: 4544      Segment ID: 2500
ATM address: 47.0091810000000000000000000.00000CA01662.01
data forward: vcd 14, 2 members, 0 packets, 0 unicasts

lecid  vcd    pkts   ATM Address
   1   11        0 47.0091810000000000000000000.00000CA01660.01
   2   17        0 47.0091810000000000000000000.00000CA04960.01
```

Table 22-2 describes significant fields in the sample displays of the **show lane bus** command.

Table 22-2 *show lane bus Field Descriptions*

Field	Description
LE BUS ATM2/0.2	Interface and subinterface for which information is displayed.
ELAN name	Name of the emulated LAN for this BUS.
Admin	Administrative state, either up or down.
State	Status of this LANE BUS. Possible states include down and operational.
type	Type of emulated LAN.
Max Frame Size	Maximum frame size (in bytes) on the emulated LAN.
Segment ID	The emulated LAN's ring number. This field appears only for Token Ring LANE.
ATM address	ATM address of this LANE BUS.
data forward	Virtual channel descriptor of the Data Forward VCC, the number of LANE clients attached to the VCC, and the number of packets transmitted on the VCC.
lecid	Identifier assigned to each LANE client on the Data Forward VCC.
vcd	Virtual channel descriptor used to reach the LANE client.
pkts	Number of packets sent by the BUS to the LANE client.
ATM Address	ATM address of the LANE client.

show lane client

To display detailed LANE information for all the LANE clients configured on an interface or any of its subinterfaces, on a specified subinterface, or on an emulated LAN, use the **show lane client** EXEC command.

> **show lane client** [**interface atm** *slot/port*[*.subinterface-number*] | **name** *elan-name*] [**brief**] (for the AIP on the Cisco 7500 series routers; for the ATM port adapter on theCisco 7200 series)

> **show lane client** [**interface atm** *slot/port-adapter/port*[*.subinterface-number*] | **name** *elan-name*] [**brief**] (for the ATM port adapter on the Cisco 7500 series routers)

> **show lane client** [**interface atm** *number*[*.subinterface-number*] | **name** *elan-name*] [**brief**] (for the Cisco 4500 and 4700 routers)

Syntax	Description
interface atm *slot/port*	(Optional) ATM interface slot and port for the following:
	• AIP on the Cisco 7500 series routers.
	• ATM port adapter on the Cisco 7200 series routers.
interface atm *slot/port-adapter/port*	(Optional) ATM interface slot, port adapter, and port number for the ATM port adapter on the Cisco 7500 series routers.
interface atm *number*	(Optional) ATM interface number for the NPM on the Cisco 4500 or 4700 routers.
.subinterface-number	(Optional) Subinterface number.
name *elan-name*	(Optional) Name of emulated LAN. The maximum length of the name is 32 characters.
brief	(Optional) Displays the brief subset of available information.

Command Mode
EXEC

Usage Guidelines
This command first appeared in Cisco IOS Release 11.0.

Sample Displays
The following is sample output from the **show lane client** command for an Ethernet emulated LAN:

```
Router# show lane client

LE Client ATM2/0.2  ELAN name: elan2  Admin: up  State: operational
Client ID: 1                LEC up for 11 minutes 49 seconds
Join Attempt: 1
HW Address: 0000.0ca0.5b40   Type: ethernet          Max Frame Size: 1516

ATM Address: 39.020304050607080910111213.00000CA05B40.02

 VCD  rxFrames  txFrames  Type     ATM Address
   0         0         0  configure 39.020304050607080910111213.00000CA05B43.00
```

```
55          1          4   direct     39.020304050607080910111213.00000CA05B41.02
56          6          0   distribute 39.020304050607080910111213.00000CA05B41.02
59          0          1   send       39.020304050607080910111213.00000CA05B42.02
60          3          0   forward    39.020304050607080910111213.00000CA05B42.02
84          3          5   data       39.020304050607080910111213.00602F557940.02
```

The following is sample output from the **show lane client** command for a Token Ring LANE:

```
Router# show lane client

LE Client ATM4/0.1  ELAN name: elan1  Admin: up  State: operational
Client ID: 1                  LEC up for 2 hours 26 minutes 3 seconds
Join Attempt: 3
HW Address: 0060.4770.4180   Type: token ring      Max Frame Size: 4544
Ring:100    Bridge:2        ELAN Segment ID: 2048
ATM Address: 39.020304050607080910111213.006047704180.01

VCD  rxFrames  txFrames  Type       ATM Address
  0         0         0  configure  39.020304050607080910111213.006047704183.00
 10         1         3  direct     39.020304050607080910111213.006047704181.01
 11         2         0  distribute 39.020304050607080910111213.006047704181.01
 14         0         0  send       39.020304050607080910111213.006047704182.01
 15         0         0  forward    39.020304050607080910111213.006047704182.01
```

Table 22-3 describes significant fields in the sample displays of the **show lane client** command.

Table 22-3   *show lane client Field Descriptions*

Field	Description
LE Client ATM2/0.2	Interface and subinterface of this client.
ELAN name	Name of the emulated LAN.
Admin	Administrative state; either up or down.
State	Status of this LANE client. Possible states include initialState, lecsConnect, configure, join, busConnect, and operational.
Client ID	The LAN emulation 2-byte Client ID assigned by the LAN emulation server.
Join Attempt	The number of attempts before successfully joining the emulated LAN.
HW Address	MAC address of this LANE client.
Type	Type of emulated LAN.
Max Frame Size	Maximum frame size (in bytes) on the emulated LAN.
Ring	The ring number for the client. This field only appears for Token Ring LANE.
Bridge	The bridge number for the client. This field only appears for Token Ring LANE.

Table 22-3   *show lane client* Field Descriptions

Field	Description
ELAN Segment ID	The ring number for the emulated LAN. This field only appears for Token Ring LANE.
ATM Address	ATM address of this LANE client.
VCD	Virtual channel descriptor for each of the VCCs established for this LANE client.
rxFrames	Number of frames received.
txFrames	Number of frames transmitted.
Type	Type of VCC. The Configure Direct VCC is shown in this display as *configure*. The Control Direct VCC is shown as *direct*; the Control Distribute VCC is shown as *distribute*. The Multicast Send VCC and Multicast Forward VC are shown as *send* and *forward*, respectively. The Data Direct VCC is shown as *data*.
ATM Address	ATM address of the LANE component at the other end of this VCC.

show lane config

To display global LANE information for the configuration server configured on an interface, use the **show lane config** EXEC command.

> **show lane config** [**interface atm** *slot*/**0**]
> (for the AIP on the Cisco 7500 series routers; for the ATM port adapter on the Cisco 7200 series)

> **show lane config** [**interface atm** *slot/port-adapter*/**0**]
> (for the ATM port adapter on the Cisco 7500 series routers)

> **show lane config** [**interface atm** *number*]
> (for the Cisco 4500 and 4700 routers)

Syntax	Description
interface atm *slot*/**0**	(Optional) ATM interface slot and port for the following: • AIP on the Cisco 7500 series routers. • ATM port adapter on the Cisco 7200 series routers.

Syntax	Description
interface atm *slot/port-adapter/***0**	(Optional) ATM interface slot, port adapter, and port number for the ATM port adapter on the Cisco 7500 series routers.
interface atm *number*	(Optional) ATM interface number for the NPM on the Cisco 4500 or 4700 routers.

Command Mode

EXEC

Usage Guidelines

This command first appeared in Cisco IOS Release 11.0.

Sample Displays

The following is sample output from the **show lane config** command for an Ethernet emulated LAN:

```
Router# show lane config

LE Config Server ATM2/0 config table: cisco_eng
Admin: up  State: operational
LECS Mastership State: active master
list of global LECS addresses (30 seconds to update):
39.020304050607080910111213.00000CA05B43.00  <-------- me
ATM Address of this LECS: 39.020304050607080910111213.00000CA05B43.00 (auto)
 vcd  rxCnt  txCnt  callingParty
  50     2      2  39.020304050607080910111213.00000CA05B41.02 LES elan2 0 active
cumulative total number of unrecognized packets received so far: 0
cumulative total number of config requests received so far: 30
cumulative total number of config failures so far: 12
    cause of last failure: no configuration
    culprit for the last failure: 39.020304050607080910111213.00602F557940.01
```

The following example shows sample output of the **show lane config** command for TR-LANE:

```
Router# show lane config

LE Config Server ATM4/0 config table: eng
Admin: up  State: operational
LECS Mastership State: active master
list of global LECS addresses (40 seconds to update):
39.020304050607080910111213.006047704183.00  <-------- me
ATM Address of this LECS: 39.020304050607080910111213.006047704183.00 (auto)
 vcd  rxCnt  txCnt  callingParty
```

```
     7       1       1   39.020304050607080910111213.006047704181.01 LES elan1 0 active
cumulative total number of unrecognized packets received so far: 0
cumulative total number of config requests received so far: 2
cumulative total number of config failures so far: 0
```

Table 22-4 describes significant fields in the sample displays of the **show lane config** command.

Table 22-4   *show lane config Field Descriptions*

Field	Description
LE Config Server	Major interface on which the LANE configuration server is configured.
config table	Name of the database associated with the LANE configuration server.
Admin	Administrative state, either up or down.
State	State of the configuration server: down or operational. If down, the reasons field indicates why it is down. The reasons include the following: *NO-config-table*, *NO-nsap-address*, and *NO-interface-up*.
LECS Mastership state	Mastership state of the configuration server. If you have configured simple server redundancy, the configuration server with the lowest index is the active LECS.
list of global LECS addresses	List of LECS addresses.
40 seconds to update	Amount of time until the next update.
<-------- me	ATM address of this configuration server.
ATM Address of this LECS	ATM address of the active configuration server.
auto	Method of ATM address assignment for the configuration server. In this example, the address is assigned by the automatic method.
vcd	Virtual circuit descriptor that uniquely identifies the configure VCC.
rxCnt	Number of packets received.
txCnt	Number of packets transmitted.
callingParty	ATM NSAP address of the LANE component that is connected to the LECS. *elan1* indicates the emulated LAN name, *0* indicates the priority number, and *active* indicates that the server is active.

show lane database

To display the configuration server's database, use the **show lane database** EXEC command.

> **show lane database** [*database-name*]

Syntax

database-name

Description

(Optional) Specific database name.

Command Mode

EXEC

Usage Guidelines

This command first appeared in Cisco IOS Release 11.0.

By default, this command displays the LANE configuration server information displayed by the **show lane config** command.

If no database name is specified, this command shows all databases.

Sample Displays

The following is sample output of the **show lane database** command for an Ethernet LANE:

```
Router# show lane database

LANE Config Server database table 'engandmkt' bound to interface/s: ATM1/0
default elan: none
elan 'eng': restricted
  server 45.000001415555121f.yyyy.zzzz.0800.200c.1001.01 (prio 0) active
  LEC MAC   0800.200c.1100
  LEC NSAP 45.000001415555121f.yyyy.zzzz.0800.200c.1000.01
  LEC NSAP 45.000001415555124f.yyyy.zzzz.0800.200c.1300.01
elan 'mkt':
  server 45.000001415555121f.yyyy.zzzz.0800.200c.1001.02 (prio 0) active
  LEC MAC   0800.200c.1200
  LEC NSAP 45.000001415555121f.yyyy.zzzz.0800.200c.1000.02
  LEC NSAP 45.000001415555124f.yyyy.zzzz.0800.200c.1300.02
```

The following is sample output of the **show lane database** command for a Token Ring LANE:

```
Router# show lane database

LANE Config Server database table 'eng' bound to interface/s: ATM4/0
default elan: elan1
elan 'elan1': un-restricted, local-segment-id 2048
  server 39.020304050607080910111213.006047704181.01 (prio 0) active
```

Table 22-5 describes significant fields in the sample displays of the **show lane database** command.

Table 22-5   *show lane database Field Descriptions*

Field	Description
LANE Config Server database	Name of this database and interfaces bound to it.
default elan	Default name, if one is established.
elan	Name of the emulated LAN whose data is reported in this line and the following indented lines.

Table 22-5   *show lane database* *Field Descriptions (Continued)*

Field	Description
un-restricted	Indicates whether this emulated LAN is restricted or unrestricted.
local-segment-id 2048	Ring number of the emulated LAN.
server	ATM address of the configuration server.
(prio 0) active	Priority level and simple server redundancy state of this configuration server. If you have configured simple server redundancy, the configuration server with the lowest priority will be active.
LEC MAC	MAC addresses of an individual LANE client in this emulated LAN. This display includes a separate line for every LANE client in this emulated LAN.
LEC NSAP	ATM addresses of all LANE clients in this emulated LAN.

show lane default-atm-addresses

To display the automatically assigned ATM address of each LANE component in a router or on a specified interface or subinterface, use the **show lane default-atm-addresses** EXEC command.

> **show lane default-atm-addresses** [**interface atm** *slot/port.subinterface-number*]
> (for the AIP on the Cisco 7500 series routers; for the ATM port adapter
> on the Cisco 7200 series)

> **show lane default-atm-addresses** [**interface atm** *slot/port-adapter/*
> *port.subinterface-number*] (for the ATM port adapter on the Cisco 7500
> series routers)

> **show lane default-atm-addresses** [**interface atm** *number.subinterface-number*]
> (for the Cisco 4500 and 4700 routers)

Syntax	Description
interface atm *slot/port*	(Optional) ATM interface slot and port for the following:
	• AIP on the Cisco 7500 series routers.
	• ATM port adapter on the Cisco 7200 series routers.
interface atm *slot/port-adapter/port*	(Optional) ATM interface slot, port adapter, and port number for the ATM port adapter on the Cisco 7500 series routers.
interface atm *number*	(Optional) ATM interface number for the NPM on the Cisco 4500 or 4700 routers.
.subinterface-number	(Optional) Subinterface number.

Command Mode
EXEC

Usage Guidelines

The **show lane default-atm-addresses** [**interface atm** *slot/port.subinterface-number*] command first appeared in Cisco IOS Release 11.0.

The **show lane default-atm-addresses** [**interface atm** *number.subinterface-number*] command first appeared in Cisco IOS Release 11.1.

It is not necessary to have any of the LANE components running on this router before you use this command.

Sample Display

The following is sample output of the **show lane default-atm-addresses** command for the ATM interface 1/0 when all the major LANE components are located on that interface:

```
Router# show lane default-atm-addresses interface atm1/0

interface ATM1/0:
LANE Client:        47.00000000000000000000000000.00000C304A98.**
LANE Server:        47.00000000000000000000000000.00000C304A99.**
LANE Bus:           47.00000000000000000000000000.00000C304A9A.**
LANE Config Server: 47.00000000000000000000000000.00000C304A9B.00
note: ** is the subinterface number byte in hex
```

Table 22-6 describes significant fields shown in the output of the **show lane default-atm-addresses** command.

Table 22-6 *show lane default-atm-addresses Field Descriptions*

Field	Description
interface ATM1/0:	Specified interface.
LANE Client:	ATM address of the LANE client on the interface.
LANE Server:	ATM address of the LANE server on the interface.
LANE Bus:	ATM address of the LANE BUS on the interface.
LANE Config Server:	ATM address of the LANE configuration server on the interface.

show lane le-arp

To display the LE ARP table of the LANE client configured on an interface or any of its subinterfaces, on a specified subinterface, or on an emulated LAN, use the **show lane le-arp** EXEC command.

show lane le-arp [**interface atm** *slot/port*[*.subinterface-number*] | **name** *elan-name*]
(for the AIP on the Cisco 7500 series routers; for the ATM port adapter
on the Cisco 7200 series)

show lane le-arp [**interface atm** *slot/port-adapter/port*[*.subinterface-number*] |
name *elan-name*] (for the ATM port adapter on the Cisco 7500 series routers)

show lane le-arp [**interface atm** *number*[*.subinterface-number*] | **name** *elan-name*]
(for the Cisco 4500 and 4700 routers)

Syntax / Description

Syntax	Description
interface atm *slot/port*	(Optional) ATM interface slot and port for the following:
	• AIP on the Cisco 7500 series routers.
	• ATM port adapter on the Cisco 7200 series routers.
interface atm *slot/port-adapter/port*	(Optional) ATM interface slot, port adapter, and port number for the ATM port adapter on the Cisco 7500 series routers.
interface atm *number*	(Optional) ATM interface number for the NPM on the Cisco 4500 or 4700 routers.
.subinterface-number	(Optional) Subinterface number.
name *elan-name*	(Optional) Name of emulated LAN. The maximum length of the name is 32 characters.

Command Mode

EXEC

Usage Guidelines

This command first appeared in Cisco IOS Release 11.0.

Sample Displays

The following is sample output of the **show lane le-arp** command for an Ethernet LANE client:

```
Router# show lane le-arp

Hardware Addr   ATM Address                                      VCD  Interface
0000.0c15.a2b5  39.000000000000000000000000.00000C15A2B5.01  39   ATM1/0.1
0000.0c15.f3e5  39.000000000000000000000000.00000C15F3E5.01  25*  ATM1/0.1
```

The following is sample output of the **show lane le-arp** command for a Token Ring LANE client:

```
Router# show lane le-arp

Ring Bridge     ATM Address                                         VCD  Interface
512   6         39.020304050607080910111213.00602F557940.01  47   ATM2/0.1
```

Table 22-7 describes significant fields shown in the displays of the **show lane le-arp** command.

Table 22-7   *show lane le-arp Field Descriptions*

Field	Description
Hardware Addr	MAC address, in dotted hexadecimal notation, assigned to the LANE component at the other end of this VCD.
Ring	Route descriptor segment number for the LANE component.
Bridge	Bridge number for the LANE component.
ATM Address	ATM address of the LANE component at the other end of this VCD.
VCD	Virtual circuit descriptor.
Interface	Interface or subinterface used to reach the specified component.

show lane server

To display global information for the LANE server configured on an interface, on any of its subinterfaces, on a specified subinterface, or on an emulated LAN, use the **show lane server** EXEC command.

> **show lane server** [**interface atm** *slot/port*[*.subinterface-number*] | **name** *elan-name*] [**brief**] (for the AIP on the Cisco 7500 series routers; for the ATM port adapter on the Cisco 7200 series)
>
> **show lane server** [**interface atm** *slot/port-adapter/port*[*.subinterface-number*] | **name** *elan-name*] [**brief**] (for the ATM port adapter on the Cisco 7500 series routers)
>
> **show lane server** [**interface atm** *number*[*.subinterface-number*] | **name** *elan-name*] [**brief**] (for the Cisco 4500 and 4700 routers)

Syntax	Description
interface atm *slot/port*	(Optional) ATM interface slot and port for the following:

- AIP on the Cisco 7500 series routers.

- ATM port adapter on the Cisco 7200 series routers.

interface atm *slot/port-adapter/port*   (Optional) ATM interface slot, port adapter, and port number for the ATM port adapter on the Cisco 7500 series routers.

interface atm *number*   (Optional) ATM interface number for the NPM on the Cisco 4500 or 4700 routers.

.subinterface-number   (Optional) Subinterface number.

name *elan-name*   (Optional) Name of emulated LAN. The maximum length of the name is 32 characters.

brief   (Optional) Keyword used to display the brief subset of available information.

Command Mode
EXEC

Usage Guidelines
This command first appeared in Cisco IOS Release 11.0.

Sample Displays
The following is sample output from the **show lane server** command for an Ethernet emulated LAN:

```
Router# show lane server

LE Server ATM2/0.2  ELAN name: elan2  Admin: up  State: operational
type: ethernet        Max Frame Size: 1516
ATM address: 39.020304050607080910111213.00000CA05B41.02
LECS used: 39.020304050607080910111213.00000CA05B43.00 connected, vcd 51
control distribute: vcd 57, 2 members, 2 packets

proxy/ (ST: Init, Conn, Waiting, Adding, Joined, Operational, Reject, Term)
lecid ST vcd   pkts Hardware Addr  ATM Address
    1  0   54     2 0000.0ca0.5b40 39.020304050607080910111213.00000CA05B40.02
    2  0   81     2 0060.2f55.7940 39.020304050607080910111213.00602F557940.02
```

The following is sample output from the **show lane server** command for a Token Ring emulated LAN:

```
Router# show lane server

LE Server ATM3/0.1  ELAN name: anubis  Admin: up  State: operational
type: token ring        Max Frame Size: 4544      Segment ID: 2500
ATM address: 47.00918100000000000000000.00000CA01661.01
LECS used: 47.00918100000000000000000.00000CA01663.00 connected, vcd 6
control distribute: vcd 10, 2 members, 4 packets
proxy/ (ST: Init, Conn, Waiting, Adding, Joined, Operational, Reject, Term)
lecid ST vcd   pkts Hardware Addr  ATM Address
    1  0   7      3 400.1           47.00918100000000000000000.00000CA01660.01
                    0000.0ca0.1660 47.00918100000000000000000.00000CA01660.01
    2  0  16      3 300.1           47.00918100000000000000000.00000CA04960.01
                    0000.0ca0.4960 47.00918100000000000000000.00000CA04960.01
```

Table 22-8 describes significant fields shown in the displays of the **show lane server** command.

Table 22-8   *show lane server Field Descriptions*

Field	Description
LE Server ATM2/0.2	Interface and subinterface of this server.
ELAN name	Name of the emulated LAN.
Admin	Administrative state, either up or down.
State	Status of this LANE server. Possible states for a LANE server include *down*, *waiting_ILMI*, *waiting_listen*, *up_not_registered*, *operational*, and *terminating*.
type	Type of emulated LAN.
Max Frame Size	Maximum frame size (in bytes) on this type of emulated LAN.
Segment ID	The emulated LAN's ring number. This field appears only for Token Ring LANE.
ATM address	ATM address of this LANE server.
LECS used	ATM address of the LANE configuration server being used. This line also shows the current state of the connection between the LANE server and the LANE configuration server and the VCD of the circuit connecting them.
control distribute	Virtual circuit descriptor of the Control Distribute VCC.
proxy	Status of the LANE client at the other end of the Control Distribute VCC.
lecid	Identifier for the LANE client at the other end of the Control Distribute VCC.
ST	Status of the LANE client at the other end of the Control Distribute VCC. Possible states are *Init*, *Conn*, *Waiting*, *Adding*, *Joined*, *Operational*, *Reject*, and *Term*.

Table 22-8   *show lane server* Field Descriptions

Field	Description
vcd	VCD used to reach the LANE client.
pkts	Number of packets sent by the LANE server on the Control Distribute VCC to the LANE client.
Hardware Addr	The top number in this column is the router-descriptor, while the second number is the MAC-layer address of the LANE client.
ATM Address	ATM address of the LANE client.

Part
VII

Command Reference

CHAPTER 23

MPOA Overview

This chapter describes the Multiprotocol over ATM (MPOA) feature, which is supported in Cisco IOS Release 11.3(4)WA4(6) and later.

MPOA enables the fast routing of internetwork-layer packets across a nonbroadcast multiaccess (NBMA) network. MPOA replaces multihop routing with point-to-point routing using a direct VCC between ingress and egress edge devices or hosts. An ingress edge device or host is defined as the point at which an inbound flow enters the MPOA system; an egress edge device or host is defined as the point at which an outbound flow exits the MPOA system.

The following components are required for an MPOA network:

- MPOA client (MPC)

- MPOA server (MPS)

- Catalyst 5000 series ATM module

- LANE

- NHRP

An MPC identifies packets sent to an MPS, establishes a shortcut VCC to the egress MPC, and then routes these packets directly over the shortcut VCC. An MPC can be a router or a Catalyst 5000 series ATM module. An MPS can be a router or a Catalyst 5000 series RSM/VIP2 with an ATM interface.

NOTE Because the RSM/VIP2 can also be used as a router, all references to *router* in this document refer to both a router and the RSM/VIP2 with an ATM interface.

MPOA provides the following benefits:

- Eliminates multiple router hops between the source and the destination points of the ATM cloud by establishing shortcuts for IP packets and other protocol packets.

- Frees the router for other tasks by reducing IP traffic.

- Provides backward compatibility as an ATM network by building upon LANE, and can be implemented using both MPOA and LANE-only devices.

Platforms

The MPOA feature is supported on the following platforms running Cisco IOS software Release 11.3(4)WA4(6) or later:

● Catalyst 5000 series ATM module

● Catalyst 5000 series RSM/VIP2 with an ATM interface

● Cisco 7200 and 7500 series routers

● Cisco 4500 and 4700 series routers

How MPOA Works

In an NBMA network, intersubnet routing involves forwarding packets hop-by-hop through intermediate routers. MPOA can increase performance and reduce latencies by identifying the edge devices, establishing a direct VCC between the ingress and egress edge devices, and forwarding Layer 3 packets directly over this shortcut VCC, bypassing the intermediate routers. An MPC provides the direct VCCs between the edge devices or hosts whenever possible and forwards Layer 3 packets over these shortcut VCCs. The MPCs must be used with MPSs resident on routers. (See Figure 23-1.)

Figure 23-1 *MPOA Message Flow Between MPCs and MPSs*

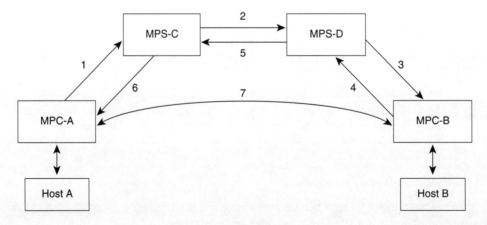

The sequence of events shown in Figure 23-1 is summarized as follows:

> **Step 1** MPOA resolution request sent from MPC-A to MPS-C
>
> **Step 2** NHRP resolution request sent from MPS-C to MPS-D
>
> **Step 3** MPOA cache-imposition request sent from MPS-D to MPC-B
>
> **Step 4** MPOA cache-imposition reply sent from MPC-B to MPS-D

Step 5	NHRP resolution reply sent from MPS-D to MPS-C
Step 6	MPOA resolution reply sent from MPS-C to MPC-A
Step 7	Shortcut VCC established

Table 23-1 lists and defines the MPOA terms used in Figure 23-1.

Table 23-1   *MPOA Terms*

MPOA Term	Definition
MPOA resolution request	A request from an MPC to resolve a destination protocol address to an ATM address to establish a shortcut VCC to the egress device.
NHRP resolution request	An MPOA resolution request which has been converted to an NHRP resolution request.
MPOA cache-imposition request	A request from an egress MPS to an egress MPC providing the MAC rewrite information for a destination protocol address.
MPOA cache-imposition reply	A reply from an egress MPC acknowledging an MPOA cache-imposition request.
NHRP resolution reply	An NHRP resolution reply that eventually will be converted to an MPOA resolution reply.
MPOA resolution reply	A reply from the ingress MPS resolving a protocol address to an ATM address.
Shortcut VCC	The path between MPCs over which Layer 3 packets are sent.

Traffic Flow

Figure Figure 23-1 shows how MPOA messages flow from Host A to Host B. In this figure, an MPC (MPC-A) residing on a host or edge device detects a packet flow to a destination IP address (Host B) and sends an MPOA resolution request. An MPS (MPS-C) residing on a router converts the MPOA resolution request to an NHRP resolution request and passes it to the neighboring MPS/NHS (MPS-D) on the routed path. When the NHRP resolution request reaches the egress point, the MPS (MPS-D) on that router sends an MPOA cache-imposition request to MPC-B. MPC-B acknowledges the request with a cache-imposition reply and adds a tag that allows the originator of the MPOA resolution request to receive the ATM address of MPC-B. As a result, the shortcut VCC between the edge MPCs (MPC-A and MPC-B) is set up.

When traffic flows from Host A to Host B, MPC-A is the ingress MPC and MPC-B is the egress MPC. The ingress MPC contains a cache entry for Host B with the ATM address of the egress MPC. The ingress MPC switches packets destined to Host B on the shortcut VCC with the appropriate tag received in the MPOA resolution reply. Packets traversing through the shortcut VCC do not have any DLL headers. The egress MPC contains a cache entry that associates the IP address of Host B and the ATM address of the ingress MPC to a DLL header. When the egress MPC switches an IP packet through a shortcut path to Host B, it appears to have come from the egress router.

Interaction with LANE

An MPOA functional network must have at least one MPS, one or more MPCs, and zero or more intermediate routers implementing NHRP servers. The MPSs and MPCs use LANE control frames to discover each other's presence in the LANE network.

CAUTION   For MPOA to work properly, you must first create an emulated LAN identifier for each emulated LAN. Use the **lane config database** or the **lane server-bus** ATM LANE commands to create emulated LAN identifiers.

An MPC/MPS can serve as one or more LECs. The LEC can be associated with any MPC/MPS in the router or Catalyst 5000 series switch. A LEC can be attached to both an MPC and an MPS simultaneously.

Figure 23-2 shows the relationships between MPC/MPS and LECs.

Figure 23-2   *MPC-LEC and MPS-LEC Relationships*

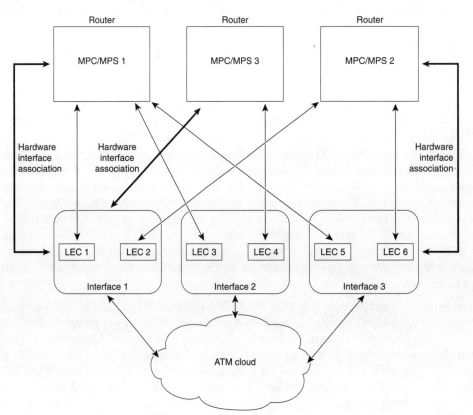

Configuring an MPC/MPS

To configure an MPC/MPS, perform the following tasks:

- Define a name for the MPC/MPS

- Attach the MPC/MPS to a major interface. This task serves two purposes:

    — Assigns an ATM address to the MPC/MPS.

    — Identifies an endpoint for initiating and terminating MPOA virtual circuits.

- Bind the MPC/MPS to multiple LECs.

Multiple MPCs/MPSs can run on the same physical interface, each corresponding to a different control ATM address. Once an MPC/MPS is attached to a single interface for its control traffic, it cannot be attached to another interface unless you break the first attachment. The MPC/MPS is attached to a subinterface 0 of the interface.

In Figure 23-2, MPC/MPS 1 is attached to interface 1; MPC/MPS 1 can only use interface 1 to set up its control VCs. MPC/MPS 2 is attached to interface 3; MPC/MPS 2 can only use interface 3 to set up its control VCs.

NOTE   An MPC/MPS can be attached to a single hardware interface only.

More than one MPC/MPS can be attached to the same interface. MPC/MPS 3 and MPC/MPS 1 are both attached to interface 1, although they get different control addresses. Any LEC running on any subinterface of a hardware interface can be bound to any MPC/MPS. However, once a LEC is bound to a particular MPC/MPS, it cannot be bound to another MPC/MPS.

NOTE   Once a LEC has been bound to an MPC/MPS, you must unbind the LEC from the first MPC/MPS before binding it to another MPC/MPS. Typically, you will not need to configure more than one MPS in a router.

Ensure that the hardware interface attached to an MPC/MPS is directly reachable through the ATM network by all the LECs that are bound to it.

NOTE   If any of the LECs reside on a different (unreachable) ATM network from the one to which the hardware interface is connected, MPOA will not operate properly.

CHAPTER 24

Configuring the MPOA Client

This chapter describes the required and optional tasks for configuring the MPOA client. For a complete description of the MPC commands used in this chapter, refer to Chapter 26, "MPOA Commands." For documentation of other commands that appear in this chapter, you can search online at www.cisco.com.

The MPC functionality involves ingress/egress cache management, data-plane and control-plane VCC management, MPOA frame processing, and participation in MPOA protocol and MPOA flow detection.

How MPC Works

The MPC software module implements the functionality of the MPC in compliance with the ATM Forum MPOA specification. An MPC identifies packets sent to an MPOA-capable router over the NBMA network and establishes a shortcut VCC to the egress MPC, if possible. The MPC then routes these packets directly over this shortcut VCC, bypassing the intermediate routers and enabling the fast routing of internetwork-layer packets across an NBMA network. The Catalyst 5000 series switch can be designated as an MPC. If the Catalyst 5000 series switch is configured with an RSM/VIP2 (with an ATM interface) it can be configured as an MPC or an MPS.

A router is usually designated as an MPS, but can also be designated as an MPC. MPC on the router is primarily meant to provide router-initiated and router-terminated shortcuts for non-NBMA networks. For this reason, MPC information in this publication primarily refers to the Catalyst 5000 series switch, and MPS information refers to the router or the RSM/VIP2 with an ATM interface in a Catalyst 5000 series switch.

MPC Configuration Task List

To configure an MPC on your network, perform the following tasks. Only the first two tasks are required; the remaining two tasks are optional:

- Configuring the ELAN ID
- Configuring the MPC
- Configuring the MPC Variables
- Monitoring and Maintaining the MPC

Configuring the ELAN ID

For MPOA to work properly, a LEC must belong to an emulated LAN that has a defined ELAN ID. To obtain an ELAN ID, use either of the following commands:

NOTE   To configure an MPC on a Catalyst 5000 series ATM module, establish connection with the ATM module, enter privileged mode, and then enter configuration mode.

Command	Purpose
name *elan-name* **elan-id** *id*	Defines an ELAN ID for the LEC (in LANE database configuration mode).
lane server-bus ethernet *elan-name* [**elan-id** *id*]	Configures the LEC with the ELAN ID (in interface configuration mode).

CAUTION   If an ELAN ID is supplied, make sure both commands use the same *elan-id* value.

Configuring the MPC

To configure an MPC on your network, use the following commands in the appropriate configuration modes:

Step	Command	Purpose
1	**mpoa client config name** *mpc-name* (In global configuration mode)	Defines an MPC with a specified name.
2	**interface atm** {*mod-num/port-num* | *number*} (In interface configuration mode)	Specifies the ATM interface with which the MPC is to be associated.
3	**mpoa client name** *mpc-name* (In interface configuration mode)	Attaches an MPC to the ATM interface.
4	**interface** *atm-num.sub-interface-num* **mul** (In interface configuration mode)	Specifies the ATM interface that contains the LEC to which you will bind the MPC.
5	**lane client mpoa client name** *mpc-name* (In interface configuration mode)	Binds a LANE client to the specified MPC.

Repeat Steps 4 and 5 for every LEC to be served by the MPC/MPS.

NOTE	In Step 4, if you do not specify the **mul** keyword when entering a subinterface number, the CLI does not accept the command.

Configuring the MPC Variables

An MPC has to be defined with a specified name before you can change its variables.

To change the variables for an MPC, use the following commands, beginning in MPC configuration mode:

Step	Command	Purpose
1	**mpoa client config name** *mps-name*	Defines an MPC with the specified name.
2	**atm-address** *atm-address*	(Optional) Specifies the control ATM address that the MPC should use (when it is associated with a hardware interface).
3	**shortcut-frame-count** *count*	(Optional) Specifies the maximum number of times a packet can be routed to the default router within shortcut-frame time before an MPOA resolution request is sent.
4	**shortcut-frame-time** *time*	(Optional) Sets the shortcut-setup frame time for the MPC.

Monitoring and Maintaining the MPC

To monitor and maintain the configuration of an MPC, use any of these commands in EXEC mode:

Command	Purpose
show mpoa client [**name** *mpc-name*]	Displays information about a specified MPC or all MPCs.
show mpoa client [**name** *mpc-name*] **cache** [**ingress** | **egress**] [**ip-addr** *ip-addr*]	Displays ingress and egress cache entries associated with an MPC.
show mpoa client [**name** *mpc-name*] **statistics**	Displays all the statistics collected by an MPC.
clear mpoa client [**name** *mpc-name*] **cache** [**ingress** | **egress**] [**ip-addr** *ip-addr*]	Clears cache entries.
show mpoa client [**name** *mpc-name*] [*remote-device*]	Displays all the MPOA devices that this MPC has learned.
show mpoa default-atm-address	Displays the default ATM addresses for the MPC.

MPC Configuration Examples

This section contains an example of the commands needed to configure an MPC. The lines beginning with exclamation points (!) are comments explaining the command shown on the subsequent line. Figure 24-1 shows an example of how you can configure your system to use MPOA.

Figure 24-1 *An Example of an MPOA Configuration*

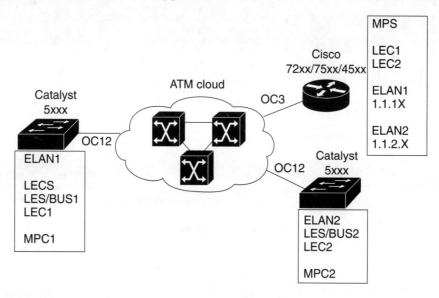

The following example configures the MPC and attaches the MPC to a hardware interface:

```
! Define the MPC "MYMPC"
 mpoa client config name MYMPC
! Leave everything as default
 exit
! Specify the ATM interface to which the MPC is attached
 interface ATM 1/0
! Attach MPC MYMPC to the HW interface
 mpoa client name MYMPC
! Specify the ATM interface that contains the LEC to which you will bind the MPC
 interface atm 1/0.1
! Bind a LANE client to the specified MPC
 lane client mpoa client name MYMPC
! Go back up to global config mode
 exit
```

The following example shows a typical configuration file for the first MPC:

```
Current configuration:
!
version 11.3
```

```
! Go to LANE database config mode
 exit
lane database mpoa-test
hostname mpc-1
! Define the ELAN ID and ATM address
name elan1 server-atm-address 47.00918100000000613E5A2F01.006070174821.01
name elan1 elan-id 101
name elan2 server-atm-address 47.00918100000000613E5A2F01.006070174821.02
name elan2 elan-id 102
! Define the MPC "mpc-1"
mpoa client config name mpc-1
        interface Ethernet0
! Go back up to global config mode
 exit
! Specify the ATM interface to which the MPC is attached
interface ATM0
        atm pvc 1 0 5 qsaal
        atm pvc 2 0 16 ilmi
        lane config auto-config-atm-address
        lane config database mpoa-test
! Attach MPC mpc-1 to the HW interface
        mpoa client name mpc-1
! Specify the ATM interface that contains the LEC to which you will bind the MPC
interface ATM0.1 multipoint
        lane server-bus ethernet elan1
! Bind a LANE client to the specified MPC
        lane client mpoa client name mpc-1
        lane client ethernet 1 elan1
! Go back up to global config mode
 exit
```

The following example shows a typical configuration file for the second MPC:

```
Current configuration:
!
version 11.3
hostname mpc-2
! Go back up to global config mode
 exit
! Define the MPC "mpc-2"
mpoa client config name mpc-2
! Specify the ATM interface to which the MPC is attached
interface ATM0
        atm pvc 1 0 5 qsaal
        atm pvc 2 0 16 ilmi
        mpoa client name mpc-2
! Specify the ATM interface that contains the LEC to which you will bind the MPC
interface ATM0.1 multipoint
        lane server-bus ethernet elan2
        lane client mpoa client name mpc-2
        lane client ethernet 2 elan2
! Go back up to global config mode
 exit
```

CHAPTER 25

Configuring the MPOA Server

This chapter describes the required and optional tasks for configuring the MPOA server. For a complete description of the MPS commands used in this chapter, refer to Chapter 26, "MPOA Commands." For documentation of other commands that appear in this chapter, you can search online at www.cisco.com.

The MPS supplies the forwarding information used by the MPCs. The MPS responds with the information after receiving a query from a client. To support the query and response functions, MPOA has adopted the NHRP. The MPS on the router can also terminate shortcuts.

How MPS Works

The MPS software module implements the functionality of the MPS in compliance with the ATM Forum MPOA specification. The following sections describe the functions of MPS:

- MPS-NHRP Routing Interaction
- Shortcut Domains

MPS-NHRP Routing Interaction

MPS has to interact with the NHRP module in the router to smoothly propagate MPOA/NHRP packets end-to-end. MPOA frames are identical to NHRP frames except for some specific op-codes and extensions for MPOA.

The following process explains the interaction of MPS and NHRP:

Step 1   MPS converts MPOA resolution requests to NHRP requests and sends it either to the next hop MPS or to the next-hop server (NHS), depending on the configuration. MPS searches for the next-hop routing information to determine the interface and sends the packet with correct encapsulation to an MPS or an NHS.

Step 2   NHS sends resolution requests to MPS when the next hop is on a LANE cloud or when NHS is unsure of the packet destination. MPS may do further processing, such as prompt NHS to terminate the request or throw away the packet.

Step 3   NHS sends resolution replies to MPS when the next hop interface is LANE or when the replies terminate in the router. Then MPS sends an MPOA resolution reply to the MPC.

Shortcut Domains

Within a router, it is possible to permit shortcuts between one group of LECs and deny it between some other groups of LECs. Cisco introduces a notion of network ID associated with an MPS. By default, all the MPSs in a router get a network ID of 1.

If the administrator wants to segregate traffic, then MPSs can be given different network IDs, in effect preventing shortcuts between LECs served by different MPSs. This can be configured in the definition of an MPS database.

If a router has both MPS and NHRP configured, then the same network ID is required to facilitate requests, replies, and shortcuts across the MPS and NHRP. The interface-specific NHRP command (**ip nhrp network-id**) must be the same for an MPS; otherwise, there will be a disjointed network.

MPS Configuration Task List

To configure an MPS on your network, perform the following tasks. Only the first two tasks are required; the remaining two tasks are optional:

● Configuring the ELAN ID

● Configuring the MPS

● Configuring the MPS Variables

● Monitoring and Maintaining the MPS

Configuring the ELAN ID

For MPOA to work properly, a LANE client must have an ELAN ID for all ELANs represented by the LANE clients. To configure an ELAN ID, use either of the following commands in LANE database configuration mode or in interface configuration mode when starting up the LES for that emulated LAN:

Command	Purpose
name *elan-name* **elan-id** *id*	Configures the ELAN ID in the LECS database to participate in MPOA.
lane server-bus {ethernet | tokenring} *elan-name* [**elan-id** *id*]	Configures the LES with the ELAN ID to participate in MPOA.

CAUTION	If an ELAN ID is supplied by both commands, make sure that the ELAN ID matches in both.

Configuring the MPS

To configure an MPS, use the following commands. The MPS starts functioning only after it is attached to a specific hardware interface:

Step	Command	Purpose
1	**mpoa server config name** *mps-name*	In global configuration mode, it defines an MPS with the specified name.
2	**interface atm** {*slot/port*	*number*}	Specifies the ATM interface to attach the MPS.
3	**mpoa server name** *mps-name*	In interface configuration mode, it attaches the MPS to the ATM interface.
4	**interface atm** {*slot/port.subinterface-number*	*number.subinterface-number*}	Specifies the ATM interface to bind the MPS to a LEC.
5	**lane client mpoa server name** *mps-name*	In subinterface configuration mode, it binds a LANE client to the specified MPS.

Configuring the MPS Variables

An MPS has to be defined with a specified name before you can change the MPS variables specific to that MPS. To change MPS variables specific only to a particular MPS, use the following commands, starting in MPS configuration mode:

Step	Command	Purpose
1	**mpoa server config name** *mps-name*	Defines an MPS with the specified name.
2	**atm-address** *atm-address*	(Optional) Specifies the control ATM address that the MPS should use (when it is associated with a hardware interface).
3	**network-id** *id*	(Optional) Specifies the network ID of the MPS.
4	**keepalive-time** *time*	(Optional) Specifies the keepalive time value for the MPS-p1 variable of the MPS.
5	**holding-time** *time*	(Optional) Specifies the holding time value for the MPS-p7 variable of the MPS.

Monitoring and Maintaining the MPS

To monitor and maintain the configuration of an MPS, use the following commands in EXEC mode:

Command	Purpose
show mpoa default-atm-addresses	Displays default ATM addresses for an MPS.

Command	Purpose
show mpoa server [**name** *mps-name*]	Displays information about a specified server or all servers, depending on the specified name of the required server.
show mpoa server [**name** *mps-name*] **cache** [**ingress** | **egress**] [**ip-address** *ip-address*]	Displays ingress and egress cache entries associated with a server.
show mpoa server [**name** *mps-name*] **statistics**	Displays all the statistics collected by a server, including the ingress and egress cache entry creations, deletions, and failures.
clear mpoa server [**name** *mps-name*] **cache** [**ingress** | **egress**] [**ip-addr** *ip-addr*]	Clears cache entries.
mpoa server name *mps-name* **trigger ip-address** *ip-address* [**mpc-address** *mpc-address*]	Originates an MPOA trigger for the specified IP address to the specified client. If a client is not specified, the MPOA is triggered to all the clients.

MPS Configuration Examples

This section contains an example of the commands needed to configure an MPS. The lines beginning with exclamation points (!) are comments explaining the command shown on the following line. Figure 25-1 shows an example of how you can configure your system to utilize MPOA.

Figure 25-1 *Example of an MPOA Configuration*

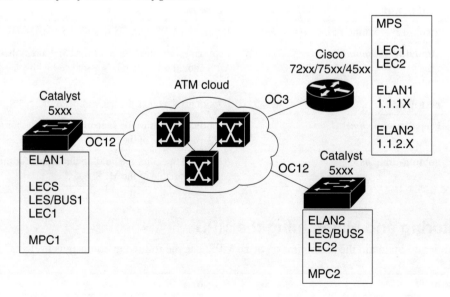

MPS Configuration Examples

The following example configures the MPS and attaches the MPS to a hardware interface:

```
! Define the MPS "MYMPS"
 mpoa server config name MYMPS
! Leave everything as default
 exit
! Enter into interface config mode
 interface ATM 1/0
! Attach MPS MYMPS to the HW interface
 mpoa server name MYMPS
! Go back up to global config mode
 exit
```

The following example shows a typical MPS configuration file:

```
version 11.3
hostname MPS
! Define the MPS "mps"
mpoa server config name mps
! Specify the ATM interface to which the MPS is attached
interface ATM4/0
    no ip address
    atm pvc 1 0 5 qsaal
    atm pvc 2 0 16 ilmi
    lane config auto-config-atm-address
    mpoa server name mps
! Specify the ATM interface that contains the LEC to which you will bind the MPS
interface ATM4/0.1 multipoint
    ip address 1.1.1.2 255.255.255.0
    lane client mpoa server name mps
    lane client ethernet elan1
interface ATM4/0.2 multipoint
    ip address 1.1.2.1 255.255.255.0
    lane client mpoa server name mps
    lane client ethernet elan2
end
```

CHAPTER 26

MPOA Commands

This chapter describes the commands available to configure and maintain MPOA. For MPOA configuration information and examples, refer to Chapter 23, "MPOA Overview," Chapter 24, "Configuring the MPOA Client," and Chapter 25, "Configuring the MPOA Server."

atm-address

To override the control ATM address of an MPC or MPS, use the **atm-address** interface configuration command. Use the **no** form of this command to revert to the default address.

> **atm-address** *atm-address*
> **no atm-address**

Syntax	Description
atm-address	Control ATM address.

Default

The default is an auto-generated ATM address.

Command Mode

Interface configuration

Usage Guidelines

This command first appeared in Cisco IOS Release 11.3(3a)WA4(5).

This command specifies the control ATM address that an MPC or MPS should use when it comes up; that is, when it is associated with a hardware interface.

The **atm-address** command overrides the default operational control address of the MPC or MPS. When this address is deleted (using the **no** form of the command), the MPC or MPS uses an auto-generated address as its control address.

Example

The following example specifies the ATM address for an MPC:

```
mpoa-client-config#atm-address 47.0091810000000061705b7701.00400BFF0011.00
```

The following example specifies the ATM address for an MPS:

```
mpoa-server-config#atm-address 47.0091810000000061705C2B01.00E034553024.00
```

clear mpoa client cache

Use the **clear mpoa client cache** EXEC command to clear the ingress and egress cache entries of one or all MPCs.

clear mpoa client [**name** *mpc-name*] **cache** [**ingress** | **egress**] [**ip-address** *ip-address*]

Syntax	Description
name *mpc-name*	(Optional) Specifies the name of the MPC with the specified name.
ingress	(Optional) Clears ingress cache entries associated with the MPC.
egress	(Optional) Clears egress cache entries associated with the MPC.
ip-address *ip-address*	(Optional) Clears matching cache entries with the specified IP address.

Defaults

The system defaults are

- All MPC cache entries are cleared.
- Both caches are cleared.
- Entries matching only the specified destination IP address are cleared.

Command Mode

EXEC

Usage Guideline

This command first appeared in Cisco IOS Release 11.3(3a)WA4(5).

Example

The following example clears the ingress and egress cache entries for the MPC named *ip_mpc*:

```
ATM#clear mpoa client name ip_mpc cache
```

Related Commands

You can search online at www.cisco.com to find documentation of related commands.

show mpoa client cache

clear mpoa server cache

To clear the ingress and egress cache entries, use the **clear mpoa server cache** EXEC command.

> **clear mpoa server** [**name** *mps-name*] **cache** [**ingress** | **egress**] [**ip-address** *ip-address*]

Syntax	Description
name *mps-name*	(Optional) Specifies the name of the MPS. If **name** *mps-name* is omitted, the command applies to all servers.
ingress	(Optional) Clears ingress cache entries associated with a server.
egress	(Optional) Clears egress cache entries associated with a server.
ip-address *ip-address*	(Optional) Clears matching cache entries with the specified IP address. If **ip-address** *ip-address* is omitted, the command clears all entries.

Command Mode

EXEC

Usage Guidelines

This command first appeared in Cisco IOS Release 11.3(3a)WA4(5).

This command clears cache entries.

Example

The following example clears all cache entries:

```
ATM#clear mpoa server cache
```

Related Commands

You can search online at www.cisco.com to find documentation of related commands.

show mpoa server cache

holding-time

To specify the holding time value for the MPS-p7 variable of an MPS, use the **holding-time** MPS configuration command. To revert to the default value, use the **no** form of this command.

> **holding-time** *time*
> **no holding-time** *time*

Syntax	Description
time	Specifies the holding time value in seconds.

Default

The default holding time is 1,200 seconds (20 minutes).

Command Mode

MPS configuration

Usage Guidelines

This command first appeared in Cisco IOS Release 11.3(3a)WA4(5).

Example

The following example sets the holding time to 600 seconds (10 minutes):

```
holding-time 600
```

keepalive-time

To specify the keepalive time value for the MPS-p1 variable of an MPS, use the **keepalive-time** MPS configuration command. To revert to the default value, use the **no** form of this command.

keepalive-time *time*
no keepalive-time *time*

Syntax

Description

time   Specifies the keepalive time value in seconds.

Default
The default keepalive time is 10 seconds.

Command Mode
MPS configuration

Usage Guidelines
This command first appeared in Cisco IOS Release 11.3(3a)WA4(5).

Example
The following example sets the keepalive time to 25 seconds:

```
keepalive-time 25
```

lane client mpoa client name
Use the **lane client mpoa client name** interface configuration command to bind a LEC to the named MPC. Use the **no** form of this command to unbind the named MPC from a LEC.

lane client mpoa client name *mpc-name*
no lane client mpoa client name *mpc-name*

Syntax   Description

mpc-name   Name of the specific MPC.

Default
No LEC is bound to a named MPC.

Command Mode
Interface configuration

Usage Guidelines

This command first appeared in Cisco IOS Release 11.3(3a)WA4(5).

When you enter this command, the named MPC is bound to a LEC. The named MPC must exist before this command is accepted. If you enter this command before a LEC is configured (not necessarily running), a warning message is issued.

Example

The following example binds a LEC on a subinterface to the MPC:

```
ATM (config-subif)#lane client mpoa client name ip_mpc
```

lane client mpoa server name

To bind a LEC with the named MPS, use the **lane client mpoa server name** interface configuration command. To unbind the server, use the **no** form of this command.

> **lane client mpoa server name** *mps-name*
> **no lane client mpoa server name** *mps-name*

Syntax	Description
mps-name	Name of the specific MPOA server.

Default

No LEC is bound to a named MPS.

Command Mode

Interface configuration

Usage Guidelines

This command first appeared in Cisco IOS Release 11.3(3a)WA4(5).

This command binds a LEC to the named MPS. The specified MPS must exist before this command is accepted. If this command is entered when a LEC is not already configured (not necessarily running), a warning message is issued.

Example

The following example binds a LANE client with the MPS named *MYMPS*:

```
ATM (config-subif)#lane client mpoa server name MYMPS
```

mpoa client config name

Use the **mpoa client config name** global configuration command to define an MPC with a specified name. Use the **no** form of this command to delete the MPC.

> **mpoa client config name** *mpc-name*
> **no mpoa client config name** *mpc-name*

Syntax   Description

mpc-name   Specifies the name of an MPC.

Default

This command has no default setting.

Command Mode

Global configuration

Usage Guidelines

This command first appeared in Cisco IOS Release 11.3(3a)WA4(5).

When you configure/create an MPC, you automatically enter the MPC configuration mode. From here, you can enter subcommands to define or change MPC variables specific only to this MPC. Note that the MPC is not functional until it is attached to a hardware interface.

Example

The following example creates or modifies the MPC named *ip_mpc*:

```
ATM> enable
ATM#configure
ATM (config)#mpoa client config name ip_mpc
mpoa-client-config#
```

Related Commands

You can search online at www.cisco.com to find documentation of related commands.

atm-address
shortcut-frame-count
shortcut-frame-time

mpoa client name

Use the **mpoa client name** interface configuration command to attach an MPC to a major ATM interface. Use the **no** form of this command to break the attachment.

> **mpoa client name** *mpc-name*
> **no mpoa client name** *mpc-name*

Syntax   Description

mpc-name   Specifies the name of an MPC.

Default

No MPC is attached to an ATM interface.

Command Mode

Interface configuration

Usage Guidelines

This command first appeared in Cisco IOS Release 11.3(3a)WA4(5).

The **mpoa client name** command provides an interface to the MPC through which the MPC can set up and receive calls.

When you enter this command on a major interface that is up and operational, the named MPC becomes operational. Once the MPC is fully operational, it can register its ATM address.

Example

The following example attaches the MPC *ip_mpc* to an interface:

```
mpoa-client-config#interface atm 1/0
config-if#mpoa client name ip_mpc
```

mpoa server config name

To define an MPS with the specified name, use the **mpoa server config name** global configuration command. To delete an MPS, use the **no** form of this command.

> **mpoa server config name** *mps-name*
> **no mpoa server config name** *mps-name*

Syntax

Syntax	Description
mps-name	Name of the MPOA server.

Default

No MPS is defined.

Command Mode

Global configuration

Usage Guidelines

This command first appeared in Cisco IOS Release 11.3(3a)WA4(5).

This command defines an MPS with the specified name. The MPS does not actually start functioning until it is attached to a specific hardware interface. Once that attachment is complete, the MPS starts functioning. When you configure/create an MPS, you automatically enter the MPS configuration mode.

You can define the MPS variables specific to an MPS only after that MPS is defined with a specified name. After this command is entered, further commands may be used to change MPS variables that are specific only to this MPS.

Example

The following example defines the MPS named *MYMPS*:

```
ATM (config)#mpoa server config name MYMPS
```

mpoa server name

To attach an MPS to a major ATM interface, use the **mpoa server name** interface configuration command. To break the attachment, use the **no** form of this command.

> **mpoa server name** *mps-name*
> **no mpoa server name** *mps-name*

Syntax   Description

mps-name   Name of the MPOA server.

Default

No MPS is attached to an ATM interface.

Command Mode

Interface configuration

Usage Guidelines

This command first appeared in Cisco IOS Release 11.3(3a)WA4(5).

This command attaches an MPS to a specific (major) interface. At this point, the MPS has the capability to obtain its auto-generated ATM address and an interface through which it can communicate to the neighboring MPOA devices. Only when an MPS is both defined globally and attached to an interface is it considered to be operational. Although multiple different servers may share the same hardware interface, an MPS can be attached to only a single interface at any one time. Note that the specified MPS must already have been defined when this command is entered.

Example

The following example attaches the MPS named *MYMPS* to an ATM interface:

```
ATM (config)#mpoa server name MYMPS
```

mpoa server name trigger ip-address

To originate an MPOA trigger for the specified IP address to the specified MPOA client from the specified MPS, use the **mpoa server name trigger ip-address** EXEC command.

> **mpoa server name** *mps-name* **trigger ip-address** *ip address* [**mpc-address** *mpc-address*]

Syntax	Description
mps-name	Specifies the name of the MPOA server.
ip address	Specifies the IP address.
mpc-address *mpc-address*	(Optional) Specifies the MPC address to which the trigger should be sent. If the address is not specified, a trigger is sent to all clients.

Command Mode

EXEC

Part
VII

Command Reference

Usage Guidelines

This command first appeared in Cisco IOS Release 11.3(3a)WA4(5).

This command sends an MPOA trigger for the specified IP address to the specified MPOA client from the specified MPOA server. If an MPOA client is not specified, it is triggered to all MPOA clients.

Example

The following example sends an MPOA trigger for the specified IP address 128.9.0.7 to all known MPOA clients from the MPOA server named *MYMPS*:

```
Router#mpoa server name MYMPS trigger ip-address 128.9.0.7
```

network-id

To specify the network ID of an MPS, use the **network-id** MPS configuration command. To revert to the default value, use the **no** form of this command.

> **network-id** *id*
> **no network-id**

Syntax	Description
id	Specifies the network ID of the MPOA server.

Default

The default value for **network-id** is 1.

Command Mode

MPS configuration

Usage Guidelines

This command first appeared in Cisco IOS Release 11.3(3a)WA4(5).

Specifies the network ID of this MPS. This value is used similarly to how the NHRP network ID is used: It is for partitioning NBMA clouds artificially by administration.

Example

The following example sets the network ID to 5:

```
network-id 5
```

shortcut-frame-count

Use the **shortcut-frame-count** MPC configuration command to specify the maximum number of times a packet can be routed to the default router within shortcut-frame time before an MPOA resolution request is sent. Use the **no** form of this command to restore the default shortcut-setup frame count value.

> **shortcut-frame-count** *count*
> **no shortcut-frame-count**

Syntax	Description
count	Shortcut-setup frame count.

Default

The default is 10 frames.

Command Mode

MPC configuration mode

Usage Guideline

This command first appeared in Cisco IOS Release 11.3(3a)WA4(5).

Example

The following example sets the shortcut-setup frame count to 5 for the MPC:

```
mpoa-client-config#shortcut-frame-count 5
```

Related Commands

You can search online at www.cisco.com to find documentation of related commands.

atm-address
mpoa client config name
shortcut-frame-time

shortcut-frame-time

Use the **shortcut-frame-time** MPC configuration command to set the shortcut-setup frame time (in seconds) for the MPC. Use the **no** form of this command to restore the default shortcut-setup frame-time value.

> **shortcut-frame-time** *time*
> **no shortcut-frame-time**

Syntax	Description
time	(Optional) Shortcut-setup frame time in seconds.

Default

The default is 1 second.

Command Mode

MPC configuration

Usage Guideline

This command first appeared in Cisco IOS Release 11.3(3a)WA4(5).

Example

The following example sets the shortcut-setup frame time to 7 for the MPC:

```
mpoa-client-config#shortcut-frame-time 7
```

Related Commands

You can search online at www.cisco.com to find documentation of related commands.

atm-address
mpoa client config name
shortcut-frame-count

show mpoa client

Use the **show mpoa client** EXEC command to display a summary of information regarding one or all MPCs.

> **show mpoa client** [**name** *mpc-name*] [**brief**]

Syntax	Description
name *mpc-name*	(Optional) Name of the MPC with the specified name.
brief	(Optional) Output limit of the command.

Command Mode

EXEC

Usage Guidelines

This command first appeared in Cisco IOS Release 11.3(3a)WA4(5).

If you omit the **name** keyword, the command displays information for all MPCs.

Sample Display

The following is sample output from the **show mpoa client** command:

```
ATM#show mpoa client name ip_mpc brief
MPC Name: ip_mpc, Interface: ATM1/0, State: Up
MPC actual operating address: 47.00918100000000613E5A2F01.0010A6943825.00
Shortcut-Setup Count: 1, Shortcut-Setup Time: 1
Lane clients bound to MPC ip_mpc: ATM1/0.1
Discovered MPS neighbors                    kp-alv    vcd      rxPkts     txPkts
47.00918100000000613E5A2F01.006070174824.00    59      30          28          2
Remote Devices known                                  vcd      rxPkts     txPkts
47.00918100000000613E5A2F01.00000C5A0C5D.00           35           0         10
```

Table 26-1 describes the fields shown in the display of the **show mpoa client** command.

Table 26–1   *show mpoa client Field Descriptions*

Field	Description
MPC Name	Name specified for the MPC.
Interface	Interface to which the MPC is attached.
State	Current state of the MPC.
MPC actual operating address	ATM address of the MPC.
Shortcut-Setup Count	Current number specified by the **shortcut-frame-count** command.
Shortcut-Setup Time	Current value specified by the **shortcut-frame-time** command.
Lane clients bound to MPC ip_mpc	List of LANE clients currently bound to MPC *ip_mpc*.
Discovered MPS neighbors	List of learned MPS addresses.
kp-alv	Number of seconds until the next keepalive message should be received.
vcd	Number that identifies the virtual circuit.
rxPkts	Number of packets received from the learned MPS.
txPkts	Number of packets transmitted to the learned MPS.
Remote Devices known	List of other devices (typically other MPCs) not in this emulated LAN.
vcd	Number that identifies the virtual circuit to that MPC.
rxPkts	Number of packets received from the learned remote device.
txPkts	Number of packets transmitted to the learned remote device.

Part
VII

Command Reference

Related Commands

You can search online at www.cisco.com to find documentation of related commands.

clear mpoa client name

show mpoa client cache

Use the **show mpoa client cache** EXEC command to display the ingress or egress cache entries matching the IP addresses for the MPCs.

> **show mpoa client** [**name** *mpc-name*] **cache** [**ingress** | **egress**] [**ip-address** *ip-address*]

Syntax	Description
name *mpc-name*	(Optional) Name of the MPC with the specified name.
ingress	(Optional) Displays ingress cache entries associated with an MPC.
egress	(Optional) Displays egress cache entries associated with an MPC.
ip-address *ip-address*	(Optional) Displays cache entries that match the specified IP address.

Command Mode

EXEC

Usage Guideline

This command first appeared in Cisco IOS Release 11.3(3a)WA4(5).

Sample Display

The following is sample output from the **show mpoa client cache** command for a specific MPC:

```
ATM#show mpoa client ip_mpc cache
MPC Name: ip-mpc, Interface: ATM1/0, State: Up
MPC actual operating address: 47.00918100000000613E5A2F01.0010A6943825.00
Shortcut-Setup Count: 1, Shortcut-Setup Time: 1
Number of Ingress cache entries: 1
MPC Ingress Cache Information:
Dst IP addr    State   vcd Expires Egress MPC Atm address
20.20.20.1     RSVLD   35   11:38 47.00918100000000613E5A2F01.00000C5A0C5D.00
Number of Egress cache entries: 1
MPC Egress Cache Information:
Dst IP addr       Dst MAC        Src MAC      MPSid  Elan Expires  CacheId  Tag
10.10.10.1     0000.0c5a.0c58 0060.7017.4820     9     2   11:55        1    1
```

Table 26-2 describes the fields shown in the display of the **show mpoa client cache** command.

Table 26-2 *show mpoa client cache Field Descriptions*

Field	Description
MPC Name	Name specified for the MPC.
Interface	Interface to which the MPC is attached.
State	Current state of the MPC (up or down).
MPC actual operating address	ATM address of the MPC.

Table 26-2 *show mpoa client cache Field Descriptions (Continued)*

Field	Description
Shortcut-Setup Count	Current number specified by the **shortcut-frame-count** command.
Number of Ingress cache entries	Number of entries in the ingress cache.
MPC Ingress Cache Information:	
Dst IP addr	IP address of the destination.
State	State of the ingress cache entry[1].
vcd	Number that identifies the virtual circuit.
Expires	Time in minutes/seconds until the ingress cache entry expires.
Egress MPC ATM address	ATM address of the egress MPC.
Number of Egress cache entries	Number of entries in the egress cache.
MPC Egress Cache Information:	
Dst IP addr	IP address of the destination.
Dst MAC	MAC address of the destination.
Src MAC	MAC address of the source.
MPSid	Unique number representing the egress MPS.
Elan	Emulated LAN identifier of the emulated LAN serving this destination IP address.
Expires	Time in minutes/seconds until the egress cache entry expires.
CacheID	Cache identifier.
Tag	Tag identifier.

1. Valid states are *initialized, trigger, refresh, hold_down, resolved,* and *suspended.*

show mpoa client statistics

Use the **show mpoa client statistics** EXEC command to display all the statistics collected by an MPC.

show mpoa client [**name** *mpc-name*] **statistics**

Syntax	Description
name *mpc-name*	(Optional) Specifies the name of the MPC.

Command Mode
EXEC

Usage Guidelines
This command first appeared in Cisco IOS Release 11.3(3a)WA4(5).

This command displays all the statistics collected by an MPC.

Sample Display
The following is sample output from the **show mpoa client statistics** command for the MPC *ip_mpc*:

```
ATM#show mpoa client name ip_mpc statistics
MPC Name: ip_mpc, Interface: ATM1/0, State: Up
MPC actual operating address: 47.00918100000000613E5A2F01.0010A6943825.00
Shortcut-Setup Count: 1, Shortcut-Setup Time: 1

                              Transmitted      Received
MPOA Resolution Requests           2              0
MPOA Resolution Replies            0              2
MPOA Cache Imposition Requests     0              0
MPOA Cache Imposition Replies      0              0
MPOA Cache Purge Requests          0              0
MPOA Cache Purge Replies           0              0
MPOA Trigger Request               0              0
NHRP Purge Requests                0              0


Invalid MPOA Data Packets Received: 0
```

show mpoa default-atm-addresses

Use **show mpoa default-atm-addresses** EXEC command to display the default ATM addresses for the MPC.

>   **show mpoa default-atm-addresses**

Syntax Description
This command has no arguments or keywords.

Command Mode

EXEC

Usage Guideline

This command first appeared in Cisco IOS Release 11.3(3a)WA4(5).

Sample Displays

The following is sample output from the **show mpoa default-atm-addresses** command when the switch prefix is *not* available:

```
ATM#show mpoa default-atm-addresses
interface ATM1/0:
MPOA Server: ...006070174824.**
MPOA Client: ...006070174825.**
note: ** is the MPS/MPC instance number in hex

interface ATM2/0:
MPOA Server: ...006070174844.**
MPOA Client: ...006070174845.**
note: ** is the MPS/MPC instance number in hex
```

The following is sample shows output from the **show mpoa default-atm-addresses** command when the switch prefix is available:

```
ATM#show mpoa default-atm-addresses
interface ATM1/0:
MPOA Server: 47.00918100000000613E5A2F01.006070174824.**
MPOA Client: 47.00918100000000613E5A2F01.006070174825.**
note: ** is the MPS/MPC instance number in hex

interface ATM2/0:
MPOA Server: 47.10000000000000000000000000.006070174844.**
MPOA Client: 47.10000000000000000000000000.006070174845.**
note: ** is the MPS/MPC instance number in hex
```

Table 26-3 describes the fields shown in the example output of the **show mpoa default-atm-addresses** command.

Table 26-3   *show mpoa default-atm-address Field Descriptions*

Field	Description
interface ATM1/0	Specified interface.
MPOA Server	ATM address of the MPOA server on the interface.
MPOA Client	ATM address of the MPOA client on the interface.

Part VII

Command Reference

show mpoa server

To display information about any specified or all MPSs in the system depending on whether the name of the required MPS is specified, use the **show mpoa server** EXEC command.

> **show mpoa server** [**name** *mps-name*]

Syntax

name *mps-name*

Description

(Optional) Specifies the name of the MPOA server.

Command Mode

EXEC

Usage Guidelines

This command first appeared in Cisco IOS Release 11.3(3a)WA4(5).

This command displays information about any specified MPS or all MPSs in the system, depending on whether the name of the required MPS is specified. The command displays information about server configuration parameters. It also displays information about LECs that are bound to the MPOA server neighbors (both MPC and MPS).

Sample Displays

The following is sample output from the **show mpoa server** command, with a specified name:

```
Router# show mpoa server name ip_mps

MPS Name: ip_mps, MPS id: 0, Interface: ATM1/0, State: up
network-id: 1, Keepalive: 25 secs, Holding time: 1200 secs
Keepalive lifetime: 75 secs, Giveup time: 40 secs
MPS actual operating address: 47.00918100000000613E5A2F01.006070174824.00
Lane clients bound to MPS ip_mps: ATM1/0.1 ATM1/0.2
Discovered neighbors:
MPC 47.00918100000000613E5A2F01.00000C5A0C5D.00 vcds: 39(R,A)
MPC 47.00918100000000613E5A2F01.0010A6943825.00 vcds: 40(R,A)
```

Table 26-4 describes the fields shown in the display of the **show mpoa server** command.

Table 26-4 *show mpoa server Field Descriptions*

Field	Description
MPS Name	Name of the MPOA server.
MPS id	ID of the MPOA server.

Table 26-4   *show mpoa server Field Descriptions (Continued)*

Field	Description
Interface	Interface to which the MPS is attached.
State	State of the MPOA server: up or down.
network-id	Network ID used for partitioning.
Keepalive	Keepalive time value.
Holding time	Holding time value.
Keepalive lifetime	Keepalive lifetime value.
Giveup time	Minimum time to wait before giving up on a pending resolution request.
MPS actual operating address	Actual control address of this MPS.
Lane clients bound to MPS ip_mps	List of LANE clients served by the MPS.
Discovered neighbors	MPOA devices discovered by the clients bound to this MPS.

Related Commands

You can search online at www.cisco.com to find documentation of related commands.

clear mpoa server name

show mpoa server cache

To display ingress and egress cache entries associated with a server, use the **show mpoa server cache** EXEC command.

> show mpoa server [**name** *mps-name*] **cache** [**ingress** | **egress**] [**ip-address** *ip-address*]

Syntax	Description
name *mps-name*	(Optional) Specifies the name of a MPOA server.
ingress	(Optional) Displays ingress cache entries associated with a server.
egress	(Optional) Displays egress cache entries associated with a server.
ip-address *ip-address*	(Optional) Displays the entries which match the specified IP address.

Command Mode

EXEC

Usage Guidelines

This command first appeared in Cisco IOS Release 12.0.

This command displays ingress and egress cache entries associated with an MPS.

Sample Display

The following is sample output from the **show mpoa server cache** command, with a specified name:

```
Router# show mpoa server name ip_mps cache

MPS Name: ip_mps, MPS id: 0, Interface: ATM1/0, State: up
network-id: 1, Keepalive: 25 secs, Holding time: 1200 secs
Keepalive lifetime: 75 secs, Giveup time: 40 secs
MPS actual operating address: 47.00918100000000613E5A2F01.006070174824.00
Number of Ingress cache entries: 1
Ingress Cache information:
  IP address     Ingress MPC ATM Address                      Remaining Time
  20.20.20.1       47.00918100000000613E5A2F01.0010A6943825.00  19:07
Number of Egress cache entries: 1
Egress Cache information:
  Dst IP address  Ingress MPC ATM Address                     Remaining Time
  20.20.20.1       47.00918100000000613E5A2F01.0010A6943825.00  19:06
             src IP 20.20.20.2, cache Id 1
```

Table 26-5 describes the fields shown in the display of the **show mpoa server cache** command.

Table 26-5   *show mpoa server cache Field Descriptions*

Field	Description
MPS Name	Name of the MPOA server.
MPS id	ID of the MPOA server.
Interface	Interface to which the MPS is attached.
State	State of the MPOA server: up or down.
network-id	Network ID used for partitioning.
Keepalive	Keepalive time value.
Holding time	Holding time value.
Keepalive lifetime	Keepalive lifetime value.
Giveup time	Minimum time to wait before giving up on a pending resolution request.
MPS actual operating address	Actual control address of this MPS.

Table 26-5   *show mpoa server cache* *Field Descriptions (Continued)*

Field	Description
Number of Ingress cache entries	Number of entries in the ingress cache.
Ingress Cache information	Information of ingress cache.
IP address	IP address of the MPC.
Ingress MPC ATM Address	ATM address of the ingress MPC.
Remaining Time	Time for which the cache entry is valid.
Number of Egress cache entries	Number of entries in the egress cache.
Egress Cache information	Information of egress cache.
Dst IP address	IP address of the destination.
src IP	IP address of the source MPS which originated the NHRP resolution request.
cache Id	Cache identifier.

show mpoa server statistics

To display all the statistics collected by an MPS, use the **show mpoa server statistics** EXEC command.

show mpoa server [name *mps-name***] statistics**

Syntax	Description
name *mps-name*	(Optional) Specifies the name of a MPOA server.

Command Mode

EXEC

Usage Guidelines

This command first appeared in Cisco IOS Release 12.0.

This command displays all the statistics collected by an MPS. This pertains to the ingress /egress cache entry creation, deletion, and failures.

Sample Display

The following is a sample output from the **show mpoa server statistics** command, with a specified name:

```
Router# show mpoa server name ip_mps statistics

MPS Name: ip_mps, MPS id: 0, Interface: ATM1/0, State: up
network-id: 1, Keepalive: 25 secs, Holding time: 1200 secs
Keepalive lifetime: 75 secs, Giveup time: 40 secs
MPS actual operating address: 47.00918100000000613E5A2F01.006070174824.00
Opcode                           Transmitted    Received
---------------------------------------------------------------
MPOA Resolution Requests                            2
MPOA Resolution Replies               2
MPOA Cache Imposition Requests        1
MPOA Cache Imposition Replies                       1
MPOA Egress Cache Purge Requests                    0
MPOA Egress Cache Purge Replies       0
NHRP Resolution Requests              0             0
NHRP Resolution Replies               0             0
NHRP Purge Requests                   0             0
```

Table 26-6 describes the fields shown in the upper part of the display of the **show mpoa server statistics** command.

Table 26-6   *show mpoa server statistics Field Descriptions*

Field	Description
MPS Name	Name of the MPOA server.
MPS id	ID of the MPOA server.
Interface	Specified interface.
State	State of the MPOA server: up or down.
network-id	Network ID used for partitioning.
Keepalive	Keepalive time value.
Holding time	Holding time value.
Keepalive lifetime	Keepalive lifetime value.
Giveup time	Minimum time to wait before giving up on a pending resolution request.
MPS actual operating address	Actual control address of this MPS.

Numerics

A

G

H

J-K

L

M

T

W

weighted fair queuing, 8

worksheets for LANE planning, 260

X

Y-Z